UTILIZATION OF WASTE HEAT FROM POWER PLANTS

UTILIZATION OF WASTE HEAT FROM POWER PLANTS

David Rimberg

NOYES DATA CORPORATION
Park Ridge, New Jersey London, England
1974

Library of Congress Catalog Card Number: 74-82359
ISBN: 0-8155-0555-8
Printed in the United States

Published in the United States of America by
Noyes Data Corporation
Noyes Building, Park Ridge, New Jersey 07656

Foreword

This is another Noyes Data Technology Review extending from the energy field into the pollution area. It should prove to be a reliable reference on the utilization of waste heat emanating from electric power plants.

This book is based on studies conducted by industrial and engineering firms or university research teams under the auspices of various governmental agencies, e.g., The National Water Commission, Oak Ridge National Laboratory, Federal Water Pollution Control Administration, Office of Water Resources Research, Environmental Protection Agency, National Science Foundation, and the Department of Commerce.

Information and literature on this subject are scattered and difficult to pull together, yet one should have to go no further than this book to establish a sound background in this endeavor.

Advanced composition and production methods developed by Noyes Data are employed to bring our new durably bound books to you in a minimum of time. Special techniques are used to close the gap between "manuscript" and "completed book." Industrial technology is progressing so rapidly that time-honored, conventional typesetting, binding and shipping methods are no longer suitable. We have bypassed the delays in the conventional book publishing cycle and provide the user with an effective and convenient means of reviewing up-to-date information in depth.

The Table of Contents is organized in such a way as to serve as a subject index and provides easy access to the information contained in this book.

Contents and Subject Index

Introduction

Present steam power plants in the United States discharge as waste heat, energy equivalent to approximately twice their total present electrical generating capacity. Because this energy is degraded in temperature it is difficult to use. It represents a necessary, but unwanted, by-product of the energy conversion process for generating electricity.

Because of the growing quantities of waste heat discharged and the increasing ecological concern with energy growth, energy utilization and thermal discharge problems have stimulated an examination of methods for productively using energy presently wasted to the environment. In this regard, ineffective utilization of energy in building and industrial processes constitutes a major component of the national energy problem. Chapter 1 discusses possible reasons for the existence of ineffective utilization and also the means to promote efficient energy usage.

Chapter 2 assesses the cause, magnitude, and possible effects of heat discharges to water from steam electric power plants and related aspects of condenser cooling system operations. Chapter 3 is a discussion of the thermodynamics of the electric power generation cycle detailing the reasons for the inefficiencies in heat generation.

Ultimately all the available waste energy appears as low temperature heat. However, the term "utilization of waste heat" refers to the performance of useful functions with the heat before it is discharged to the environment. Within the framework of this book, by definition the temperature of the effluent water considered for subsequent use is approximately 38°C or 100°F. This is usually the cooling water which emanates from the condenser installations. Although several components of the steam power plant produce unused heat at much higher temperatures (i.e. turbine exhaust) only the applications that can use 100°F water are examined and discussed in Chapters 4 and 5. Because of this low temperature, there are relatively few applications to which serious consideration has been given: the use of heat for food production in agriculture and aquaculture, and the use of heat for wastewater treatment.

In the past the dissipation of waste heat was accomplished by wet and dry cooling towers and cooling ponds, lakes and streams. These methods are now being challenged by some sectors of society, and industry is being forced to consider their environmental impact. In this regard, Chapter 6 discusses the research needs necessary to equate the complicated interactions of the physical, engineering, biological and social aspects of the waste heat problem.

Energy Conservation
Through Effective Utilization

In two major sectors of the economy (building services and industrial processes), accounting for approximately 75 percent of total national energy consumption, energy utilization is found to be inefficient. It is estimated that in these two sectors, as much as 25 percent of the energy consumed annually by the nation as a whole may be lost through ineffective practices.

An intensive study on energy conservation through effective utilization was conducted for the National Bureau of Standards by C.A. Berg and the findings were reported in COM-73-10856 (February 2, 1973). A detailed summary of this work has been extracted and is given below.

Possible reasons for existence of ineffective utilization are considered, and possible means of improving effectiveness of utilization are discussed. Three possible levels of effort to promote effective utilization of energy are identified; one promotes effective use of present fuels in present processes; the second promotes utilization of presently unused energy sources; the third promotes more effective investment of energy in durable and maintainable products. Substantial latitude for improvement of effectiveness is shown to be realizable through technological efforts at these three levels.

PROBLEMS IN ENERGY SUPPLY

Present indications are that the U.S. demand for energy may well outstrip both power generating capacity and fuel supply. The basic problems in energy supply can be divided as follows. In the short range (1974 to 1980), the most important problem appears to be inadequate power-generating capacity. In the long range (2000 and beyond), the basic problem of energy is availability of fuel or of energy in another form such as solar or geothermal energy. In the intermediate range (1980 to 2000), the conservation of energy by means which do not damage the functioning of the economy could well be the most important consideration.

Broadly speaking, the problem of providing sufficient energy for future needs can be approached along either of two avenues: supply, or demand. However, these avenues are not independent of each other. For example, moderation of demand by curtailing industrial electrolytic processing could adversely affect development of energy supply capacity by causing shortages of electrical conductor material. Such interactions between supply and demand must be considered and evaluated in planning how to meet the overall energy needs of the nation.

This study concerns possible actions to affect the demand for energy. Here, again, there are two broad avenues of approach. On the one hand, demand for energy may be discouraged by increases in price and is characterized as the belt tightening approach. On the other hand, it is possible to maintain the output of energy-consuming processes, and at the same time, reduce energy consumption by improved efficiency. In the long run, effective actions to moderate demand will probably consist of a combination of these two approaches.

Efforts to improve efficiency are essentially technological in nature. The implementation of technological improvements in energy-consuming processes would probably require, or at least be greatly facilitated by, appropriate price, tax, loan, and regulatory policies, especially as these pertain to new building construction and industrial plant equipment.

The actions one might take to moderate demand should begin to take effect in the intermediate time range (1980 to 2000). As a matter of principle, actions proposed at this level should be reviewed for possible conflict with solutions of long-term problems of fuel and energy supply. For example, proposals for local combustion of fuels to drive power generation equipment and to provide for local utilization of reject heat frequently arise. Such proposals should be reviewed in light of the fact that centralized combustion facilities (e.g., large power plants) can operate at higher temperature and consequently at higher efficiency and also will usually be more efficiently operated and maintained. Centralized plants can thus usually provide higher combustion efficiency than can small low-cost plants appropriate for installation in individual buildings. Short-term measures to relieve demand on power generation via local combustion could, in the long run, result in poor use of fuel on a national basis.

The complementary relationship between efforts to increase energy supply and efforts to improve efficiency of utilization merit specific mention. The increasing national demand for energy reflects, in large part, influences which are not subject to immediate control, such as increasing population. There appears to be no way to meet the future needs of society without increasing the national capacity to supply energy. However, as will be shown, many of the ways in which energy is presently used to satisfy the needs of society are not particularly effective.

Large amounts of energy are allowed to leak out of the energy system at the point of consumption, and available techniques for utilization of reject heat and waste heat in energy consuming processes are seldom applied. Faced with a developing shortage of nonpolluting fuels and with the recognition that fuels of all types are a nonrenewable resource, it is appropriate to give serious attention to improving the effectiveness with which energy is used, as well as to improving the national capacity to supply energy.

If the effectiveness of energy utilization were not to be improved, then newly developed supplies of energy would be permitted to escape utilization through presently accepted leaks. This would require the development of surplus energy supply capacity, with all of its economic implications, merely to supply these leaks.

Thus, improving the effectiveness of energy utilization is seen here to be necessary both as a measure for conservation (to prevent the waste of a nonrenewable natural resource) and as a measure for economic optimization of investments in energy (to eliminate the need for surplus energy supply capacity).

Technological efforts to moderate demand through improved effectiveness of utilization will consist of a combination of short range measures, such as upgrading housing insulation or furnace performance, and long range measures including the application of thermal management techniques to industrial processes, improved building design and the institution of new technological means of improving the efficiency of energy utilization in devices. In the following section, the term leak plugging is used to characterize the technological efforts to improve present and future effectiveness of energy utilization.

PRESENT USES OF ENERGY IN THE UNITED STATES

One must consider the possibility that demands for energy may be moderated by improving the effectiveness of energy utilization, without sacrificing the expected output of present processes. To evaluate this possibility, it is essential to know (a) how much energy is consumed in various sectors of society, and (b) how much of the energy consumed could be saved by the use of more efficient practices. The term, waste energy, will be used here to mean energy which need not be used were presently available technology applied.

The possibility of further reductions of energy requirements for existing processes, through development of new technology, is an important subject to be dealt with in subsequent chapters.

Considerable data are available about energy consumption and are well summarized in a report by the Stanford Research Institute (SRI) entitled, *Patterns of Energy Consumption in the U.S.,* which was prepared for the Office of Science and Technology (January 1972). However, data required for the determination of waste energy are known for only a few sectors of the economy.

SRI data indicate that, of the total national energy consumption (NEC), 19.2 percent is used in residental building servies, 14.4 percent in commercial building services, 41.2 percent in industrial processes, and 25.2 percent is used in transportation. For commercial and residential buildings services, considerable information concerning the effectiveness of energy utilization is available; for industrial processes, only sample information concerning the effectiveness of energy utilization is available. The possibilities of improved energy utilization in transportation systems will not be considered.

POSSIBLE SAVINGS IN THE USE OF ENERGY

Energy Utilization in Buildings: Correctable Losses

Thermal Performance of Structures — The major uses for energy in buildings are space heating, air conditioning, and hot water heating, as shown in Table 1. Note that electric utility consumption has been allocated to each end use. Space heating of residences accounts for 11 percent of the total national energy consumption, while space heating of commercial occupancies represents an additional 6.9 percent of that total. Air conditioning in commercial and residential buildings represents 2.5 percent of the total national energy consumption.

The leaks which affect the effectiveness of heating and air conditioning in buildings are essentially the same, namely, inadequate insulation, excessive ventilation, high air infiltration rates in buildings, and excessive fenestration. To estimate the effectiveness of present levels of building insulation and ventilation in energy conservation in buildings, one may note that the FHA minimum property standards of 1965 permitted heat losses of 2,000 Btu/thousand cubic feet-degree day in residences. (This quantity, Btu/thousand cubic feet-degree day, represents the heating requirements of a building, relative to its size and the severity of the climate in which it serves.)

HUD Operation Breakthrough property standards of 1970 reduced this figure to 1,500 and the implemented FHA minimum property standards (1972) require heat losses to be less than 1,000 Btu/thousand cubic feet-degree day. The reduction in energy consumption implied by these standards is to be achieved largely by thermal insulation and control of air infiltration.

The 1972 FHA minimum property standards, of course, bear upon new construction. Few buildings are designed to exceed the performance levels of these standards, and therefore, one may assume that most of the residential buildings in use today may consume approximately 40 percent more energy for heating and air conditioning than they would, had they been insulated and sealed in accordance with present day minimum property standards.

In fact, in certain areas of the US, whole residential neighborhoods were built prior to the advent of either FHA or other insulation standards, and consist of buildings with little or no insulation and with very high air infiltration rates. These neighborhoods are, as a rule, located in areas of high heating requirements. In these neighborhoods the fuel consumption for heating is at least twice as great as would be required if insulation and infiltration control required by modern standards were applied. Since the neighborhoods in question are in areas of high heating requirements, their heat losses represent an especially significant leak in the national energy consumption system.

TABLE 1: ENERGY CONSUMPTION IN THE U.S. (1960-1968)*

Sector and End Use	Consumption 1960	Consumption 1968	Annual Rate of Growth	Percent of National Total 1960	Percent of National Total 1968
Residential					
Space heating	4,848	6,675	4.1%	11.3%	11.0%
Water heating	1,159	1,736	5.2	2.7	2.9
Cooking	556	637	1.7	1.3	1.1
Clothes drying	93	208	10.6	0.2	0.3
Refrigeration	369	692	8.2	0.9	1.1
Air conditioning	134	427	15.6	0.3	0.7
Other	809	1,241	5.5	1.9	2.1
Total	7,968	11,616	4.8	18.6	19.2
Commercial					
Space heating	3,111	4,182	3.8	7.2	6.9
Water heating	544	653	2.3	1.3	1.1
Cooking	98	139	4.5	0.2	0.2
Refrigeration	534	670	2.9	1.2	1.1
Air conditioning	576	1,113	8.6	1.3	1.8
Feedstock	734	984	3.7	1.7	1.6
Other	145	1,025	28.0	0.3	1.7
Total	5,742	8,766	5.4	13.2	14.4
Industrial					
Process steam	7,646	10,132	3.6	17.8	16.7
Electric drive	3,170	4,794	5.3	7.4	7.9
Electrolytic processes	486	705	4.8	1.1	1.2
Direct heat	5,550	6,929	2.8	12.9	11.5
Feed stock	1,370	2,202	6.1	3.2	3.6
Other	118	198	6.7	0.3	0.3
Total	18,340	24,960	3.9	42.7	41.2
Transportation					
Fuel	10,873	15,038	4.1	25.2	24.9
Raw materials	141	146	0.4	0.3	0.3
Total	11,014	15,184	4.1	25.5	25.2
National total	43,064	60,526	4.3	100.0%	100.0%

*Trillions of Btu and percent per year

Source: Patterns of Energy Consumption in the United States, Stanford Research Institute, (January 1972)

Sample field observations indicate that the state of insulation and draft sealing in existing commercial buildings is not significantly different from that in existing residences. Similar savings of space heating fuel, approximately 40 percent, may be assumed to be attainable through insulation and draft control in commercial buildings.

The infiltration of outside air accounts for approximately 25 to 50 percent of the heating and cooling requirements of individual buildings, depending upon the type of insulation installed in the buildings. Present construction practices yield infiltration rates which exceed by a factor of four the average ventilation requirements of typical buildings. Special areas in buildings, such as toilet facilities, kitchens, and conference rooms where heavy smoking may occur, have high ventilation requirements when in use. However, in present designs and in most present buildings the high ventilation rates required for use of these areas are maintained all day.

Reduction of infiltration to presently accepted levels of ventilation requirements could be assumed to yield a 10 to 20 percent reduction in national fuel requirements for space heating and air conditioning of buildings. Control of ventilation in critical areas so that high levels of ventilation were supplied only when required could provide an additional relief of up to 5 percent in total fuel requirements for space heating and air conditioning of commercial buildings.

In the above, the energy savings to be realized by bringing existing buildings up to presently recommended standards of performance have been considered. Future standards for insulation, ventilation, and infiltration may offer even greater potential for saving energy. Engineers presently studying building insulation estimate that it will be technologically and economically feasible to reduce the heating losses from buildings to approximately 700 Btu/thousand cubic feet-degree day, through use of insulation.

These estimates need to be substantiated through research and field testing; if they prove to be correct, it would be feasible to reduce total energy requirements of buildings by more than 50 percent through well-designed insulation and careful control of ventilation.

Heating and Air Conditioning Equipment – The efficiency of heating and air conditioning equipment is especially sensitive to the percent of full load at which the equipment is operated and to how well it is maintained. Heating equipment for buildings, including the home furnace as sold, is typically 75 percent efficient when run at full load.

However, the full load capacity of the equipment is seldom needed, and the equipment is most often operated intermittently in which mode it is much less efficient. In addition, small accumulations of soot on boiler surfaces and other minor unattended items of maintenance continuously reduce the efficiency of heating equipment during its life time. Taking the few field data presently available together with what is known about the effects of unattended maintenance and intermittent operation upon combustion apparatus in heating plants, it would appear reasonable to estimate that the actual efficiency of heating equipment in the field to be 50 percent or less, with units functioning at efficiencies as low as 35 percent not being uncommon.

The efficiency of air conditioning equipment varies widely. Air conditioners of the same rated output may differ by a factor of two in their power requirements. Thus, substantial savings in energy could be realized through inclusion of energy consumption in the criteria for selection of equipment, and by diligent maintenance of equipment.

Illumination – According to the energy consumption report of Southern Research Institute (January 1972), illumination of residences and commercial buildings accounts for 1.5 percent of the national energy consumption. In some cases, present levels of illumination are higher than necessary. The levels of illumination used in the U.S. office buildings exceed, by as much as a factor of two, levels of illumination used abroad, and there is no concrete evidence that the increased illumination is of any benefit to the buidling occupants.

Also, much greater use could be made of daylight in providing illumination in office buildings and residences. While the heat generated by illumination may lighten the heating load during the colder part of the year, during the time when indoor space is cooled by air conditioning systems, increased illumination imposes double energy costs upon building operations. Design techniques to permit greater use of daylight and optional use of artificial light exists, and could be more widely applied. Further research to ascertain the benefits of illumination levels to building occupants would be of great value in design to reduce excessive energy consumption in buidlings.

Hot Water Heating – Hot water heating merits special consideration in energy conservation efforts. Once the hot water is used, the water, with the energy it contains, literally goes down the drain. Furthermore, as Table 1 shows, hot water heating accounts for approximately 4 percent of the total national energy consumption. When viewed in the context of overall national energy consumption, the energy lost in hot water heating is indeed significant. A number of suggestions of ways to recapture the heat in expended hot water, using heat exchangers on drains, has been offered. Although most such proposals are in conflict with present local plumbing codes, some could be implemented through minor code modifications. However, solar hot water heaters (which are commercially available in many countries) could be employed without raising such conflicts, and could provide a relief of 2 percent or more of the total national energy requirements.

Other Areas – Other components of building operations contribute additional small amounts to national energy consumption (e.g., cooking 2.2 percent, clothes drying 0.3 percent). Although extensive data on the effectiveness of energy utilization in these areas are not available, sample observations indicate that energy is used with no greater effectiveness than in other building services. Improved design of appliances could significantly enhance the efficiency of these operations.

Summary of Energy Leakages in Buildings – The data cited indicate that present buildings consume approximately 40 percent more energy than would be required to maintain present levels of services and comfort, if the best present standards had been used in construction and if careful selection and diligent maintenance were applied to heating and air conditioning equipment. This represents a correctable loss of approximately 13.5 percent of the national energy consumption. The possible courses of action by which more effective utilization of energy in buildings might be promoted will be described below.

Thermal Effectiveness of Industrial Processes – Data pertaining to effectiveness of energy utilization in industry are less extensive than those available for building services. However, sample data indicate that, on the whole, the effectiveness of energy utilization in industry is no greater than in buildings. It would not be unreasonable to assume that energy savings of approximately 30 percent might be realized through application of present day energy conservation techniques to industrial processes. Some of the information on which this assumption is based is reviewed below.

The effectiveness with which energy is used in industry varies greatly, depending upon the nature of the industry and the size of the plant. In industries such as electric power generation and chemical refining, the nature of the industry is to convert the energy content of fuel to some more readily salable form. As a rule, optimal design of large plants in such industries is based on both initial costs and operating costs, especially fuel costs. Thus, design of piping systems to minimize pumping costs and design of pipe insulation in order to optimize the trade-off between heat loss and total costs (including maintenance) of insulation are common practices, and are representative of the consideration given to effective energy utilization in large plants for power generation and other similar industries.

In this connection, it should be noted that in large electric power plants much of the heat used to produce electricity is lost. In a typical modern power plant approximately two-thirds of the heating value of the fuel consumed must be rejected to the atmosphere. This reject heat is not waste in the present context; the rejection of this heat is required by the second law of thermodynamics. But, by siting a plant near a consumer, much of this heat could be put to use in waste processing, water purification, space heating, air conditioning, etc.

In other industries, including machine manufacturing, materials processing, and metal forming, the role of energy as an essential ingredient seems to be less clearly recognized, perhaps because energy costs have not been a major part of overall costs of operation, and the effectiveness with which individual items of plant equipment use energy has not been a major concern. Indeed, in some instances where energy costs have been taken into account in selection of plant equipment and plant design, industry has found that because of the prevailing low price of energy it has been cheaper to permit a leak of energy than to modify or replace inefficient equipment.

The assumption that this rule applies broadly appears not be be justifiable. Moreover, with the likelihood of further increases in fuel prices, industrial concern for effective utilization of energy may be expected to increase sharply. In particular, the management of small plants, to which effective energy utilization seems presently to be an item of small concern, may be expected to take a much greater interest in obtaining more effective use of the energy. It is appropriate to point out that price is not the only factor influencing industrial concern for effective energy utilization.

As examples of industrial efforts indicating the potential for energy savings in industrial operations, one may cite the following. Many large volume sales offices of gas suppliers currently have representatives assigned to advise industry how to use less fuel to conduct their present operations. The success of these recently instituted programs has yet to be measured. But, presumably the possibilities for energy savings in industry are sufficiently obvious that the gas suppliers can identify them.

Certain examples of improved equipment merit mention here. Gas fired vacuum furnaces have recently been developed for industry. Through the use of well-designed vacuum insulation, heat pipe technology, and modern heat transfer and combustion techniques, these furnaces operate with 25 percent of the total fuel consumption of previous vacuum furnaces. Other studies of the effectiveness of industrial energy utilization indicate that application of present-day heat recovery devices (e.g., heat wheels) and thermal management techniques could yield net energy savings of 30 percent or more in typical industrial operations.

Surveys of energy utilization in steel making and in related industrial operations indicate that fuel savings of as much as 39 percent could be realized in operation of certain items of equipment and that average fuel savings of 25 percent or more could be realized through application of current techniques of waste heat management and up-to-date equipment design to the industry as a whole.

Based upon the observations cited above, it is estimated that approximately 30 percent of the energy used in industrial processes could be saved through application of existing techniques, and that the use of these techniques will be economically justifiable at today's fuel prices.

The recent increases in fuel prices are expected to make energy conservation measures even more attractive to industry. The invention of more efficient devices, more efficient processes (e.g., cement making, refining, chemical processing) and especially the institution of a methodology for waste heat management in plants may be expected to yield further energy savings in industry, beyond the estimated 30 percent quoted above.

SUMMARY OF THE PROBLEM

Of the total national energy consumption, 33.6 percent goes to operate buildings and 41.2 percent goes to operate industrial processes. Present data indicate that building services consume approximately 40 percent more energy than would be required had optimal use of available technology been applied to building design and selection of equipment. In addition, sample data indicate that industrial processes consume approximately 30 percent more energy than would be required had optimal use of techniques of insulation, heat recovery, and heat management been applied. Altogether, this means that approximately one-quarter of the total energy consumed may be assumed to escape effective use, through correctable leaks at the point of utilization.

Viewed in the context of the present energy shortage, the identification of leaks which permit such a large loss of the national energy supply provokes one to ask just how such leaks came to be accepted. The answer appears to lie in the criteria of economic justification applied in energy utilization, which have been touched upon above. In energy conversion industries (e.g., power generation) economic justification of plants is based upon both initial costs and lifetime operating costs (which represent energy consumption).

However, when one comes to the point of energy utilization, economic justification of energy consuming equipment tends to be governed by initial costs. Thus, one commonly finds that high energy consumption has been designed into devices and buidlings in order to reduce initial costs. Whatever technological steps might be taken to increase the effectiveness of energy utilization will have to be coupled with steps to induce a change in the methods of economic justification of building and equipment purchases.

The purchaser should be alerted to the significance of lifetime operating costs of buildings and energy utilizing equipment, as well as initial costs. If this can be accomplished, then the use of excessive energy consumption to decrease the initial costs of energy utilizing equipment can most probably be eliminated. The technological possibilities to enhance the effectiveness of energy utilization in buildings and industrial processes can then be brought to field implementation.

MEANS TO PROMOTE EFFECTIVE ENERGY UTILIZATION

Ineffective utilization of energy is a major component of national energy problems. Here, three levels of effort are suggested at which the problem of improving effectiveness of energy utilization may be approached. Level 1 focuses upon improved effectiveness of use of present fuels in present application. Efforts at this level can be mounted through extension and coordination of present government and industry activity, and may be expected to be seen in relatively short time (approximately 2 to 5 years). Level 2 focuses upon the utilization of presently unused energy sources and fuels; efforts at this level will require new programs within government and industry, and may be expected to take effect in a longer range period (approximately 5 to 10 years).

Level 3 considers energy utilization in the broader context of the energy invested in materials and manufactured goods. Efforts at this level will require new integrated studies of the technological and economic implications of such possibilities as more efficient use of materials in construction and the design of the manufactured goods for durability and maintainability. The studies to be conducted at this level will, inevitably, bear upon the conservation of all natural resources, including energy, and will take effect over a long range period (approximately 5 to 10 years).

Important interactions exist between the efforts at these three levels. Programs for energy conservation can be constructed by implementing efforts at the three different levels simultaneously. If sufficient resources for simultaneous efforts at the three levels were not available, programs should be constructed by nesting of the three levels (e.g., Level 2 should be pursued in conjunction with Level 1; Level 3 should be pursued in conjunction with Levels 1 and 2).

Level 1—Current Fuels in Current Applications

Building Design — Two basic branches of activity are required in building design. First, design criteria and standards for energy conservation in new construction are required. However, even with present high rates of construction, approximately half of the buildings in service in the year 2000 will have been built before 1973. Thus, a technology for upgrading the thermal performance of existing structures is also required.

The principal problems of building construction in need of technological attention are insulation, draft sealing, ventilation, proper selection and maintenance of equipment, envelope design, fenestration design, and illumination.

In addition to the purely technological aspects of improved energy utilization in buildings, there are economic questions to be considered. The optimization of trade-offs between increased capital outlay and decreased operating costs over the life of a building is one such question. In this connection, reliable estimates of fuel price increases should be taken into account. Field data to confirm the economic benefits of energy conservation measures should be compiled and made known, to alert the public to the gains realizable through improved building performance. Preliminary evidence on these matters presently exists.

The evidence indicates that, at present fuel prices, increased capital expenditures on thermal upgrading of present construction (e.g., installation of additional insulation) can pay off in approximately five years. Assuming fuel prices will increase, the pay-off time may realistically be assumed to be substantially less than five years. Although this preliminary evidence is encouraging, conclusive evidence must be gathered and made known before one can expect the building purchaser to embrace the theory that it is to his financial benefit to invest in energy conserving aspects of buildings.

Thermal upgrading of existing buildings entails technological, economic, and social considerations. Materials and techniques to permit inexpensive, reliable, attractive and safe (e.g., fireproof) insulation of existing buildings require development. The economic trade-offs of thermal upgrading of existing construction require study.

In addition, some criteria for estimating the expected life of existing construction need to be developed; this involves social as well as technological considerations. Certain neighborhoods of older buildings undoubtedly should not be torn down and replaced with new construction even though modern technology could offer physical improvements.

The social effects of demolition and reconstruction would prove unacceptable. Thus, in considering the applicability of improved standards of building performance to financing of building purchases, or in judging the eligibility of neighborhoods for possible financial assistance in thermal upgrading, questions of social stability should be taken into account.

Finally, it is noted that the development of performance standards, as opposed to construction standards, by which the effectiveness of energy utilization in buildings can be judged, are required. A methodology of performance evaluation for buildings is a prerequisite for such standards. This methodology must include test methods for measuring heat transmission from the buildings, methods for determining ventilation and infiltration rates in buildings, methods for determining the effectiveness of building equipment, standard duty cycles in which to test building systems, and means of interpreting test results in terms of effectiveness of energy utilization. All of these components require development, and in turn are required in order to ascertain that field practices actually do lead to more effective use of energy.

The possibility of enhancing the effectiveness of energy utilization in buildings on a national scale, through federal standards (by FHA, VA) and through incentives (e.g., home improvement loans) is immense. Thirty-seven percent of the construction in the U.S. is either built for the Federal government or financially assisted by the Federal government and the influence of federal policy in construction extends well beyond this sector. Proper coordination of technological and economic efforts with regulatory agencies can have a powerful influence upon improving the effectiveness of energy utilization.

Industrial Processes – The first step to be taken in the study of industrial energy utilization is to determine just how efficient industrial processes really are. The sample data cited earlier indicate that, on the average, industrial processes may consume approximately 30 percent more energy than would be required to sustain present production if the processes had been designed for effective energy utilization.

Although the sample data are persuasive, further field studies are required to determine precisely what improvements in effectiveness of energy utilization are technologically feasible and economically justifiable. A national focus is required for such studies. In addition, programs to distribute information and to alert industry to the technological and economic aspects of energy utilization are required.

An additional important function which is required is the development and demonstration of methodologies of energy conservation. For example, techniques of waste heat management using heat recovery devices, application of efficient heat transfer devices such as heat pipes, and coupling between presently independent items of plant equipment, need to be demonstrated. These demonstrations would probably be most effective if they were to be conducted as industrial experiments.

Invention and innovation may be expected in industrial processes as concern for energy conservation increases. Federal laboratories may be able to contribute directly to innovation in industrial processes by developing certain generic processes useful to industry. For example, air slides using hot gases of combustion as the fluidizing media require development. If properly developed, these could enhance a number of materials processing operations. The Federal government may be able to facilitate invention and innovation in industrial processes by providing assistance in questions of patent policy and related matters.

Total Energy Systems – A Total Energy System is one in which power generation for a small complex of buildings is done locally, and the reject heat is used to provide comfort conditioning and hot water for the dwelling units. Total energy has been enthusiastically embraced by some as a means for effective fuel utilization.

Indeed, the promise of reducing the total fuel requirements of building complexes by 25 to 50 percent has been shown to be possible in principle. However, reliable field data on which public confidence in total energy could be established are scarce. Field testing of total energy systems is necessary to establish the feasibility of such systems. Moreover, technological problems bearing upon the effectiveness of total energy systems should be identified and carefully studied. Among these are the following:

Fuel Efficiency —Total energy plants are, of necessity, small installations, and as a rule, the effective temperature of combustion of a small plant is less than that in a large central power station. In large stations, one can afford to make use of heat recovery equipment and topping cycles to attain high combustion temperatures and the efficiencies pf power generation associated with them.

A topping cycle is an additional power generation plant which receives heat at the temperature of combustion, and rejects heat at the maximum temperature required by the main power plant. The topping cycle utilizes the temperature drop between the combustion chamber and the boiler of the plant to generate power.

To alleviate local thermal pollution from large plants, one might make use of bottoming cycles. In small plants the use of such equipment is not economically justifiable. By using efficient central station power generators and employing heat pumps to provide comfort conditioning, it is possible, in principle, to attain greater effectiveness of fuel consumption than by using small scale total energy plants with their limited thermal efficiencies. In addition, the combination of efficient central power generation with local heat pumps permits flexibility in the choice of fuels for power generation. This may be an important policy consideration, especially in connection with future shifts toward nuclear power. Local applications of nuclear power, as in a total energy system, seem not to be feasible for many reasons. The wisdom of long-range investments in an energy utilization system, such as total energy, which is both dependent upon fossil fuels and of limited thermal efficiency must be carefully studied even though the total energy concept appears to offer certain advantages in the short range.

On the other hand, central power stations, with their high effective temperatures of combustion and consequent high thermal efficiencies, have certain disadvantages. Thermal pollution, as was mentioned above, can be alleviated through the use of bottoming cycles which employ low temperature working fluids. But these cycles need still to be developed. In addition, high temperature combustion which yields high thermal efficiency also yields high levels of pollutants, particularly nitrogen-oxygen compounds (NO_x).

Finally, while the combination of central power and heat pumps appears extremely promising, certain environmental questions about large aerial densities of heat pumps still have to be resolved (e.g., where to obtain or reject heat without upsetting local natural conditions). Efficient, easily maintainable heat pumps have to be developed.

Finally, it is noted that there presently exists a broad spectrum of fossil fuels, some of which are low energy fuels which will not yield high temperature combustion. If it may be assumed that a sufficient quantity of the low energy fuels will be available for some reasonably long period compared with the useful life of small scale total-energy power-generation equipment, then it may be appropriate to plan total energy systems to use low energy fuels while reserving high energy fuels for central stations.

Compatibility of such total energy equipment with expansions of the national energy system through nonfossil fuel central station power plants may not pose serious problems if the supply of low energy fuel will survive the equipment. However, the supply of such fuels is not well determined at present. The present technology for utilizing such fuels in power generation needs development; and, in any event, supplies of all fossil fuels, including low energy fuels, are exhaustible. The above has not been offered to advocate either central power or total energy, but to illustrate that there are fundamental and unresolved technological questions upon which decisions as to implementation of total energy systems, and other matters of energy policy depend.

Maintainability — Small scale power units of 500 kw or less are very important to total energy planning, for these are the units which can power small complexes of 300 residence or less. However, a survey of records of performance of total energy systems in the field shows a very high rate of failure of these small scale ventures.

Moreover, studies have shown that a very large number of these failures can be ultimately attributed to faulty maintenance. To place the problem of maintenance in proper perspective one may consider that very high quality aircraft reciprocating engines provide 2,000 hours of service between major overhauls; a high quality aircraft turbojet engine provides approximately 7,000 hours of service between overhauls; a very high quality natural gas-fired reciprocating engine will provide as much as 10,000 hours service between major overhauls.

In the period between major overhauls numerous minor overhauls are commonly required. Now, the engines cited above are typical of the prime movers required in small total energy plants. At best one can expect slightly more than one year (8,600 hours) of continuous service from these prime movers before a major overhaul will be required. In the interim, several minor overhaul procedures may be required. This poses severe maintenance and management problems in operation of small total energy plants, where overhaul of the prime mover requires temporary shut down, or at least substantial reduction, of power generation.

Cost of maintenance is an especially important aspect of small scale power generation. The utilities estimate their maintenance costs at approximately one percent of sales prices. Given today's labor prices this amount is barely sufficient to support the salary and benefits of skilled mechanics to maintain such a plant.

The equipment in a total energy plant is technologically advanced (e.g., reciprocating prime movers, heat transfer apparatus, air conditioning equipment, electrical generators, switching apparatus, controls, etc.), and requires a wide range of skills for maintenance, certainly a wider range than one can hope to secure through direct employment of staff. The maintenance of small power generating units, therefore, requires careful management planning. Furthermore, technological efforts directed toward maintenance are required. For example, monitoring systems to give advanced warning of impending mechanical failure need to be developed. Also, possible management schemes wherein prime movers, or other items of equipment, might be leased from large companies having maintenance staffs of the size and breadth of competence to be effective, need to be investigated.

In general, the consideration of maintenance both in design of equipment and in its utilization needs to be developed as a component of technology. The particular case at hand, total energy systems, is but one example of the broader problem of conservation of resources through manufacture of effective, durable and maintainable equipment; this problem will be considered further below (Level 3).

At this juncture, it is merely pointed out that the development of suitable technology and methodologies (e.g., monitoring systems, management planning, determination of economic trade-offs) for establishment of maintainable plants and for determination of guidelines as to practicality of maintenance procedures, is required.

Level 2—Utilization of Unused Energy Sources

Solar Energy — The use of solar energy for space heating, air conditioning and hot water heating is one of the extremely attractive possibilities for conservation of nonrenewable energy resources. The annual incidence of solar energy on average buildings in the U.S. is six to ten times the amount required to heat the buildings. Solar energy is not only a presently unused and renewable source of energy, but during the cooling season the unused incident solar energy imposes a high load on air conditioning equipment which consumes energy from nonrenewable sources.

It would, therefore, be appropriate to mount efforts to utilize solar energy in local applications for building services. An important consideration pertaining to local application of solar energy is that most of the energy required by buildings is low temperature heat. For example, space heating requires air at approximately 80°F and water heating temperatures are commonly 135° to 150°F. These temperatures are below the reject heat temperatures of most steam power plants. In present practice, combustion of high energy fuels, such as natural gas or fuel oil is used to provide this low temperature heat. But when high energy fossil fuel is used to provide low temperature heat, its capacity to produce work, which is the precious commodity of energy, is permanently lost. Utilization of solar energy in local applications to provide the low temperature heat required by buildings, is, therefore, an important possible means for conservation.

In addition, solar energy can be employed to provide air conditioning in buildings. Absorption refrigeration equipment appears particularly attractive for solar power. Present absorption equipment is very low in efficiency, but the prospects of substantial improvements in efficiency through application of modern heat transfer technology are promising.

The chief advantages of absorption refrigeration equipment, in the present context, are that it requires very little power (mechanical compression is replaced by chemical effects so pumping for circulation is the only work required) and it is easy to maintain. These advantages weigh heavily in favor of development of solar-powered absorption machinery of improved efficiency.

Although a few technological problems require solution before one may expect solar energy in local applications to receive the enthusiastic acceptance of the consumer, many of the basic technological problems of solar energy have been solved.

By using what is known today, it would be technologically feasible and economically justifiable to apply solar energy to space heating and water heating on a national scale; approximately 50 percent of these energy requirements (representing approximately 11 percent of the national energy consumption) could be met through local application of solar energy.

The technological problems remaining in solar energy for local application are those connected with manufacturing, advances in collector design, and maintenance. The design of solar collectors for simplicity of manufacture and ease of maintenance needs to be developed. Solar collectors are currently custom-built. Modular collectors which could serve for water heating or home heating can be designed for mass manufacture, but this has yet to be done.

Maintenance-free equipment or exchangeable equipment could be produced, but this also has yet to be done. The engineering design of collectors so as to optimize collection efficiency has recently undergone significant advances, which can be incorporated into designs for maintainability and low cost. To illustrate the importance of manufacturing technology in the development of solar energy systems for local application it should be noted that in the most sophisticated solar energy devices in use today (the solid-state devices for direct conversion of sunlight to electricity in space applications) two thirds of the cost of each unit is represented by the case.

While a great deal of research has gone into improving the efficiency of the electronics and the special materials (e.g., selective absorbers) used in solar devices, rather little work has been done on designing the prosaic components of solar equipment, such as the cases cited above, so that they can be manufactured cheaply.

Yet, it is the cost of such components which largely determines the cost of the solar equipment, upon which public acceptance and economic justifiability of local solar energy applications ultimately depends. A major effort to address the technological problems of designing solar equipment for ease of manufacture and simplicity of maintenance is urgently needed.

Systems design is an additional important technological aspect of local solar energy which requires intensive study. Given any level of fuel price within the foreseeable future, the economic optimization of solar energy systems for buildings will require some booster heating or air conditioning capacity which utilizes other energy sources; the collection and storage facilities necessary to provide all building services by solar energy are simply too expensive to justify.

Economic justification of a solar energy system depends upon the trade-offs between initial capital requirements of solar devices and operating costs (e.g., fuel) of conventional equipment. It is common practice to use stock item home furnaces as booster heating equipment in solar-heated buildings; also stock item hot-water heaters and air conditioners are used as booster equipment for other services in solar powered buildings.

Of course, at present there is no alternative to installing a full size stock item home furnace as the booster in an experimental solar home. But this means that the capital costs of the solar heating system are simply an addition to the capital costs of a nonsolar building. The design of integrated solar energy systems for buildings, which use appropriately small booster equipment with sufficient capacity to boost but without surplus capacity which mostly remains idle, could have an immensely favorable influence upon the economic justification of solar energy in building services. For example, the combination of solar power with absorption refrigeration machinery, designed with reroutable circulation, could provide solar-powered air conditioning in summer and solar-powered building heating, using the absorption machine as a heat pump, in the winter. Heat could be bled from the intermediate temperature station of the absorption device for hot water heating all year.

Booster capacity in this case could be provided by low cost electrical strip heaters applied to the distillation chamber of the absorption device. In such a system, substantial capital savings appear to be possible, especially if current expectations for improved effectiveness of absorption devices are realized. In addition, a system similar to the one described above could provide for control of humidity as well as temperature, through solar energy. This would be a significant attraction to the home owner or building operator.

Although the technological problems cited above do require attention, the major obstacles to local application of solar energy are cultural and institutional. Solar collectors on roofs appear strange and impose certain constraints as to building style and orientation. Reliable data as to maintenance requirements and measured performance of solar energy equipment in actual field service are lacking.

In general, solar energy appears to the building buyer, the financier, and the building constructor as an interesting but as yet unproved idea. Testing and evaluation of field equipment can provide the information with which the institutional and cultural obstacles to the realization of these benefits might be overcome.

The use of solar energy to provide low temperature heat for buildings should be coupled with efforts to improve thermal performance of buildings themselves. Estimates in the literature pertaining to the extent to which solar energy can be used to provide space heating are generally based upon the assumption that the basic thermal design of the building will be carried out along conventional lines. For example, insulation, ventilation, infiltration, and fenestration are usually assumed to be the same as would be found in a building of conventional design and heat loss factors are taken to be the same as in conventional structures.

In a preceeding section it was pointed out that through proper thermal design of buildings the energy required to provide space heating could be reduced by 40 percent or more, compared with energy required by buildings of conventional designs. With present estimates indicating that solar energy should be capable of providing 50 percent of the space heating requirements of a conventional building, it would appear that through upgraded thermal design of the building itself it would be possible to build solar homes which derive substantially more than 50 percent of their space heating energy requirements from the sun.

The precise level of solar heating which one may hope to achieve in a suitably designed home depends upon economic trade-offs between costs of solar energy, storage facilities, and costs of additional insulation, draft sealing, double glazing, etc.

Incineration – The use of solid waste as fuel has attracted favorable attention of many technologists. Solid waste products are known to have heating values varying from one half that of fuel oil (paper) to as much as that of fuel oil (consumer plastics, tins). In addition, solid waste represents a form of unexploited fuel which will probably remain in relative abundance for some time to come.

It has been estimated that by the year 1990 the heating content of collected urban refuse could be used to generate as much as 35 thousand megawatts of electrical power. However, before the apparently rich fuel resources of solid refuse can be put to use, a number of technological problems require resolution.

These include the design of combustion plants (i.e., incinerators) to use widely varying fuels (e.g., waste paper, consumer plastics). In addition, the selection of materials to tolerate some of the highly corrosive products of combustion of solid refuse poses another important technological problem to be resolved in incineration technology. The supply of solid refuse in a given area may fluctuate seasonally.

The planning of incineration plants to operate effectively with widely varying fuel loadings and the design of electrical generating plants or district heating systems to use the heat generated by such incineration plants poses a set of demanding technological problems. Sorting of refuse according to combustion properties (e.g., separation of paper, plastic, glass and metal) poses an additional set of challenging technological problems, solutions of which are required as prerequisites to wide spread use of incineration as a source of useful heat.

In addition, the control of potentially polluting emissions from incinerator combustion chambers, which must accept fuels of widely varying character constitutes an area of required technological development.

Level 3—Energy Conservation in a Broader Context

The largest single consumer of energy in the U.S. is industry. The potential for moderating national demand for energy by improving the effectiveness of present industrial processes has already been discussed. Here we consider the possibility of conserving energy by improving the products in which the energy of processing is invested. To illustrate the questions contemplated for study, the following three specific examples are offered.

The Use of Materials as Dictated by Design Standards – In building construction, plumbing, and several other areas, the codes governing design are overly conservative for most applications. This simplifies design procedures, but leads to excessive use of materials, which in turn requires excessive use of energy. The possibility that more accurate design standards can be devised, which would permit construction, plumbing, and manufacturing operations to proceed without excessive use of material and without sacrifice of the functionality on safety of the product merits detailed study.

For example, the size of air-vent piping used in plumbing systems is chosen so that it can satisfy the needs of toilet systems in large apartment complexes or office buildings. An air-vent pipe of one-fifth the conventional size (with correspondingly smaller investment in energy of manufacture) would be completely adequate for most residences. This case and other examples of use of excessive materials should be studied in the context of conservation of energy and other natural resources.

Economic implications of shifting industrial emphasis from areas of materials production to areas of effective materials utilization, should be considered as an integral part of such a study.

Durable vs Disposable Goods – The disposable goods to which the public has become accustomed are widely recognized to constitute a drain on natural resources in general and energy in particular. The case for adoption of reusable bottles and containers, as a conservation measure, has been advocated by many.

This example, while it is but one of many one might cite, does serve to illustrate that sound arguments on both sides of the question of durable goods can be given, and that this question requires thorough objective study.

Maintainability of Machines – Manufacturing of machines is one of the largest components of U.S. industry. The possibility that the average life of machines might be extended through design for durability, careful utilization of durable materials and, especially, design for effective maintenance should be considered. If the average life of machines and other manufactured goods could be extended by, say 25 percent, then the energy requirements of the manufacturing industry might be reduced by a corresponding fraction.

The technological prerequisites for such an alteration of design and manufacturing have yet to be satisfied. Moreover, the social and economic ramifications of such a step require careful study. Nevertheless, the possibility of using more careful design to produce more durable goods in which materials, energy and in fact all natural resources are invested with greater effectiveness than in present practice appears to be an attractive possible measure for conservation of natural resources.

Generation of Waste Heat by Electric Power Plants

In the preceding chapter, it was shown that ineffective utilization of energy in buildings and industrial processes constitutes a major component of the national energy problem. Another major contributor to the energy problem is the generation of unusual heat (i.e., waste heat) from electric power generating plants. Much of the release is due to current operating efficiencies being relatively low. In this regard, this chapter will emphasize the waste heat problem associated with aqueous waste resulting from the cooling water used in electric power generating plants.

The following information on the use of water for cooling in steam electric power generation plants and the implications of such expanded use upon the preservation and enhancement of the quality of the Nation's waters, has been extracted from a study performed for the National Water Commission, and published in May 1972 as PB 210,355.

This study has identified and described (a) the cause and magnitude of the problem of waste heat discharge to water bodies, (b) the potential for reducing waste heat and/or putting it to productive uses, and (c) other effects on water bodies related to condenser cooling system operation.

THE WASTE HEAT PROBLEM

At the outset, a broad overview of the waste heat problem may be examined via the generation of electric power. All uses of energy, productive or otherwise, end up in the form of heat additions to the environment. This heat appears from both primary and secondary sources, the primary source being that which is waste heat at the time of energy conversion to man's use, and the secondary source being when the energy in one of its multiple forms is finally used to perform a particular function for man.

Of the total energy used in the United States today, approximately 80% is released directly to the atmosphere and the other 20% is released through water into the atmosphere. The following discussion addresses itself primarily to that 20% which is presently or is planned to be routed directly through water. Furthermore, since steam electric power production accounts for about 80% of all cooling water use, it is the main cause of the potential waste heat problem.

The problems, potential or existing, caused by large quantities of heat being discharged to the Nation's receiving waters are of proper public concern. An increasing population desiring a higher standard of living is creating, through the equipment and appliances it uses, a demand for electrical energy which is doubling about every 10 years. This demand is translated into the need for additional generating capacity, most of which will depend on fossil or nuclear fuel at least for the foreseeable future.

As an example of the magnitude of the problem based on present efficiencies, the heat equivalent of nearly two kilowatt-hours of energy are lost to the atmosphere for every one kilowatt-hour of electricity produced for consumption. Since water is a more efficient medium than air for condenser cooling, a great deal of this heat is being discharged into bodies of water and from there is dispersed to the atmosphere and ultimately to outer space. Considerable quantities of water are involved in this process.

Indiscriminate discharge of large quantities of low-grade heat to water poses a possible threat to the aquatic ecological system of the affected water bodies. Consumptive losses attendant to cooling processes may in some areas deplete stream flow to a point where it becomes of critical concern, especially in peak power-low flow periods. Also, there are other aspects of steam electric power plant operation which can affect water quality.

An additional problem, in terms of unwise resources use and management, is caused by (1) rigid environmental standards which largely deny the use of the waste assimilative capacity of water, and (2) the inadequate base of knowledge concerning the environmental aspects of resource use.

Furthermore, there seems to be misunderstanding on the part of those developing policy and legislation concerning the nature of heat and its dispersal in water. Heat is different from other substances which can be collected, concentrated and disposed of under controlled conditions. Heat cannot easily be concentrated in the same fashion, but must be dispersed. Heat can be dispersed to the biosphere in various ways, all of which must be considered in the establishment of policies concerning environmental quality.

The waste heat problem can be viewed from two perspectives. One perspective must deal with the near to intermediate term, a period of time during which power plants must of necessity be planned, designed, installed and operated using currently proven and available technology. This time period is expected to include much of the remaining part of the twentieth century.

The second perspective must deal with that future period when research and development efforts might have a significant impact on the means of energy generation and thus the quantity of waste heat discharge. This later period, though less predictable, should provide greater opportunity and flexibility of choice over a broader range of alternative courses of action.

WASTE ASSIMILATIVE CAPACITY

The first and simplest problem to consider when discussing waste heat is its assimilative capacity. The assimilative capacity of water is essentially involved with its ability to purify itself or to dilute wastes discharged into it, thus rendering them innocuous. The waste assimilative capacity may be defined as that quantity of waste that can be discharged to a receiving water under low flow conditions without causing adverse or deleterious effects on the water environment or affecting the designated uses of the receiving water body.

This natural process depends upon the hydrologic, biologic, and chemical characteristics of the water body, and obviously the characteristics of the waste(s) discharged.

Using bodies of water to accept waste heat provides two major benefits: (1) water as a dilution, dispersion, and dissipation medium is 4½ times more efficient than air on a weight basis and 42 times more efficient on a volume basis; and (2) the improved efficiency of the cooling process reduces the allocation of other resources required for this function (e.g., the construction of auxiliary cooling mechanisms are costly, reduce the plant efficiency, and require additional electric power to operate.)

Furthermore, discharging heat through water systems provides the capability to alter heat discharge patterns to the atmosphere, a point which may be important in the planning and design of our urban centers.

The use of water bodies as an intermediate step in heat discharge to the atmosphere also involves some potential costs. These potential costs appear as possible ecological damage to the aquatic system. The consideration of these benefits and costs involves a decision process to balance conflicting values. At the other extreme, rigid environmental standards, can deny any of the waste assimilative capacity of water.

TOTAL HEAT RELEASED

Before continuing with our discussion, it is important to point out a problem of possibly greater importance than the waste heat released to the water, that is, heat released to the atmosphere.

It is becoming increasingly apparent that man affects the climatic conditions of the earth by the release of heat and materials which can, particularly in the high heat release areas, change the opacity of the atmosphere.

Measurable climatic modifications from the release of heat and materials include, among other things, increases in mean air temperature, precipitation, and cloud cover. Actions taken to deny the use of water as an interim medium of thermal release tend to increase the total rejection of thermal energy to the atmosphere. This occurs because auxiliary cooling methods such as cooling towers reduce overall generation efficiency and therefore require higher heat inputs for equivalent electrical energy output.

It is probable that a value or level of discharge exists beyond which major meteorological impacts will result. Although no value has been established, it is generally agreed that the problem of total heat discharge would be less acute with dispersal of (1) population and (2) major primary heat sources.

WASTE HEAT AND COOLING WATER REQUIREMENTS

The type of generating facility, the plant heat rate (i.e., efficiency), the inlet water temperature, the design temperature rise across the condenser, and the type of cooling method used are the important factors involved in determining both the total quantity of heat that will be released and the cooling water requirements.

Waste Heat from Steam Electric Power Plants

Table 1 presents a comparison of the typical heat and cooling water characteristics of different types of steam electric power plants.

In contrast to the very large growth in electrical energy production, there have been only modest increases in thermal generating efficiency over the past 15 years. The thermal efficiencies of modern fossil fuel and nuclear power plants generally range between 33 and 40%. In other words, 60 to 67% of the heat generated is nonproductive and is rejected to the biosphere as waste heat.

In fossil fuel plants, a portion of this nonproductive heat is discharged directly to the air through the boiler and stack, as indicated in Table 1, but the majority is released to water through the condenser cooling system after which the heat is dissipated either directly or indirectly to the atmosphere (depending on the cooling method used). In a nuclear plant, however, there are no stack losses and in-plant losses are minimal, leaving nearly all the waste heat to be discharged through the condenser cooling system.

For example, a typical modern (40% efficient) 1,000 MW fossil fuel plant has a once-through cooling water requirement at full plant operating capacity of approximately 1,150 cfs corresponding to an inlet temperature in the range of 70° to 80°F and a water temperature rise across the condenser of 15°F.

For a similar sized conventional (33% efficient) light water reactor, 1,900 cfs would be required to provide the same rise. This simple comparison of plant types indicates that because of efficiency differences and heat release characteristics, the development of more

TABLE 1: HEAT CHARACTERISTICS OF TYPICAL STEAM ELECTRIC PLANTS
(Heat Values in Btu per kwh)

Plant Type	Thermal Efficiency (Percent)	Required Input (Heat Rate)	Total Waste Heat (Required Input Minus Heat Equivalent)[1]	Lost to Boiler, Stack[2] (etc.)	Heat Discharged to the Condenser	Cooling Water Requirement (ft^3/sec/Mw of Capacity[3])
Fossil fuel	33	10,500	7,100	1,600	5,500	1.6
Fossil fuel (recent)	40	8,600	5,200	1,300	3,900	1.15
Light water reactor	33	10,500	7,100	500	6,600	1.9
Breeder reactor[4]	42	8,200	4,800	300	4,500	1.35

[1] The heat equivalent of one kilowatt-hour of electricity (kwh) is 3,413 British thermal units (Btu).

[2] Approx. 10 to 15% x required input for fossil fuel; Approx. 3 to 5% x required input for nuclear

[3] Based on an inlet temperature in the range of 70° to 80°F and a temperature rise across the condenser of 15°F.

[4] Included for purposes of comparison.

light water reactor capacity, while advantageous from a number of other viewpoints, will only increase waste heat discharges to water and therefore increase cooling water requirements.

The Potential for Reducing Waste Heat

Although the initial cause of much waste heat discharge is the demand for more electric power, once the decision has been made that a steam electric power plant is to be constructed to meet an increment in demand, the factor which determines the quantity of waste heat produced is the efficiency of the generating facility.

Therefore, and in full consideration of the second law of thermodynamics, an increase in average plant efficiency is needed to reduce the waste heat release to the environment.

Although, as shown in Table 1 the type of generating facility affects the amount of heat lost through the condenser cooling system, efficiency is the key in attempting to reduce or eliminate any potential problems caused by discharging heat.

In order to assess the possibilities of reducing waste heat discharge from new technologies in electrical power generation, Table 2 shows an estimate of the most probable technologies in use through the end of the century, and for each method of generation, its most probable share of the nation's electrical generating capacity in the year 2000.

Also shown are the estimates of when each new generation technology will first come into practical use under both present and accelerated R&D.

It is expected that nearly two-thirds of the nation's generation capacity in year 2000 to be comprised of systems presently in widespread use, i.e., hydroelectric power (5%), fossil fuel generation (20 to 35%), and light-water nuclear reactors (30 to 40%).

The remainder would then be comprised of new technologies, i.e., nuclear breeders (10 to 15%), gas cooled reactors (10 to 20%), and other methods (5%).

TABLE 2: ELECTRICAL POWER GENERATING TECHNOLOGIES

Method of Generation	Fuel Used	Avg. Thermal Eff. of Plants Built in 1990–2000 Period	Heat Disch. to Cond. Cooling Water Btu/kwh	Date First Major Unit Could be in Operation: Present R&D Funding	Date First Major Unit Could be in Operation: Accelerated R&D Funding	Expected % of Total Capacity Year 2000
Present Systems						
Hydroelectric (conventional and pumped storage)	Water	-	-	-	-	5
Fossil fuel[1]	Coal, oil, gas	~42%	3,900	-	-	10 - 20
Shale oil, coal gasification and coal liquification (new fossil fuel)	Oil and gas	42%	3,900	1995	1985	10 - 15
Internal comb. eng.	Oil	25 - 35%	-	-	-	<1
Gas turbine	Gas, oil	20 - 30%	-	-	-	<1
Topping G.T. w/waste heat boiler	Gas, oil	40%	-	-	-	<1
Light water reactors	Uranium and thorium	~33%	6,600	-	-	30 - 40
Developing Systems for the Short Term (1970–2000)						
Gas cooled reactors	Uranium and thorium	~40%	4,800	-	-	10 - 20
Nuclear breeders	Uranium and thorium	38 - 42%	4,500	1990	1985	10 - 15
Fuel cells[2]	Partially oxidized coal, oil and gas	60%	-	1985	1980	<5

(continued)

TABLE 2: (continued)

Method of Generation	Fuel Used	Avg. Thermal Eff. of Plants Built in 1990-2000 Period	Heat Disch. to Cond. Cooling Water Btu/kwh	Date First Major Unit Could be in Operation: Present R&D Funding	Date First Major Unit Could be in Operation: Accelerated R&D Funding	Expected % of Total Capacity Year 2000
Developing Systems for the Short Term (1970-2000)						
EGD	Nat. or manuf. gas	40 - 55%	-	Never	1990	<5
MHD	Fossil or nuclear	55%	-	Never	1990	<5
MHD topping cycles	Fossil or nuclear	60%	1,700	Never	1990	<5
Geothermal	Geothermal energy	20 - 30%	-	-	-	<1
Developing Systems for the Long Term (After 2000)						
Thermoelectricity	Any heat	10 - 15%	-	Indefinite	-	0
Thermionic	Any heat	10 - 30%	-	Indefinite	-	0
Fusion	Hydrogen or helium (seawater)	75 - 95	Small	Never	2010	0
Solar	Sun's energy	14 - 25%	-	Never	1990	<1

[1] Conventional fossil fuel, excluding shale oil, coal liquefaction, and gasification.

[2] Not Central Station.

Cooling Methods

Given that a certain amount of heat will be nonproductive, the problem is how to release this heat to outer space without causing undesirable environmental impacts as it travels through the biosphere. Since water is used for condenser cooling, the problem here becomes one of dissipating the heat added to the water through some type of cooling process. A number of cooling methods are available. The most commonly used, singly or in combination, include:

(1) The once-through or run-of-river system where water is discharged directly to a receiving water without auxiliary cooling.

(2) Cooling ponds or canals where water is discharged to a holding area to be cooled, from which it may be discharged directly to a water body or recirculated through the condenser cooling system. These may include spray systems which serve to increase the heat loss per unit area.

(3) The cooling tower where water is cooled using wet or dry and mechanical or natural draft means after which it is discharged to a water body or recirculated. (Mechanical draft towers use fans to induce or force air movement. Air movement in a natural draft tower results from a chimney effect created by the difference in density between the air inside the shell and the air outside. Heat transfer in a wet tower occurs through evaporation while in a dry tower it occurs through conduction and convection.)

Although most cooling systems generally use either fresh or saline water, other sources such as sewage treatment plant effluent have been used on a limited basis. Each method has a range of costs and benefits and a different effect on water. For example, the once-through method requires a considerable withdrawal and discharges nearly all the waste heat to the receiving water with minimal consumptive use; wet cooling towers while requiring relatively little in the way of withdrawals and discharging no heat directly to the water are high consumptive users of water and are quite costly; dry cooling towers require practically no withdrawals, have practically no consumptive use, discharge no heat directly to the water, but are presently by far the most costly of cooling methods.

It is difficult to provide a useful perspective on cooling water requirements as requirements vary considerably according to range of assumptions. As an example, the *U.S. Federal Power Commission (1972)*, #14, pages I-10-16 has attempted to show how required withdrawals, consumptive use, and cooling system costs could change under a range of assumptions about possible environmental standards which the nation might choose.

EFFECTS OF STEAM ELECTRIC POWER PLANT OPERATION ON THE WATER ENVIRONMENT

Thus far we have dealt primarily with the sources of waste heat discharge. Now we turn to the effects, which form the basis for an identified problem.

Temperature is one of the most, if not the most, important single factors governing the occurrence and behavior of life. The effects of heat discharge from power plants are therefore the most pervasive of all the various effects of condenser cooling. Furthermore, as we discussed in the preceding section, the consumptive losses attendant to various cooling systems might in some areas deplete streamflow to a point where it becomes a critical concern, especially in peak power-low flow periods.

Various other aspects not directly concerned with heat but intimately related to the operation of condenser cooling systems also merit attention. Chemicals used to facilitate the cooling process, metal loss from corrosion and/or erosion, and mechanical and hydraulic effects of cooling system design and operation can have detrimental environmental effects. A number of these effects can be minimized by proper engineering design and site selection.

Effects of Temperature

The discharge of heat to a body of water can cause a number of physical, biological, and chemical effects. As water temperature is raised, the oxygen-holding capacity of the water decreases, the reaeration rate rises, density changes may cause stratification, evaporation is increased, chemical, biological, and physical reactions tend to be at a faster rate, and viscosity decreases, which reduces the sediment transporting ability of the water. These changes, of course, can be beneficial, detrimental, or insignificant depending to a great extent on the desired uses of the receiving waters.

There could be a variety of impacts. Heat increases the rate of BOD exertion which, when combined with a reduced rate of reaeration, temporarily lowers a stream's organic waste assimilation capacity. The addition of heat in the winter months could lengthen the shipping season by elimination of lock jamming and potentially shortening the period of ice cover in the shipping lanes. Discharge of heat in colder waters could also be beneficial for other uses.

The greatest potential impact of heat discharge to water, however, is on aquatic life. Discharges of large quantities of low-level heat relative to the waste assimilative capacity of the receiving waters could have severe effects on the ecosystem into which they are discharged.

A large number of studies have been and are being conducted on this subject. Although a great deal of additional in situ study must be made of specific aquatic species and their tolerance ranges and a large number of gaps must be filled in an attempt to understand cause and effect relationships, a number of basic relationships are fairly well understood.

Temperature changes have a direct effect on such things as metabolism, reproductive cycles, behavior, and digestion and respiration rates. Adverse effects can be lethal as well as sublethal. Mortality can occur because of the exceedence of maximum temperature tolerances, as well as from too rapid increases or decreases in temperature.

A shift in population structure that affects specie diversity can have a substantial impact on ecosystem stability. This can occur throughout the food web, from plankton, periphyton, filamentous algae, rooted aquatic plants, invertebrates, to fish.

Since fish are at the apex of the aquatic food pyramid, any drastic change in any part of the pyramid will be reflected in changed fish population and/or specie diversity. The more we learn about these effects the more knowledgeable will be our decision-making concerning the effects of power plants on the water environment.

Consumptive Use

A major effect of using water for condenser cooling is the loss of water to the local water regime, termed consumptive use, which accompanies any evaporative cooling process. Its magnitude is primarily dependent on the cooling system selected and therefore, especially in water-short areas, should be a significant consideration in cooling system design. An examination of a number of recently completed water resources investigations indicates that evaporative losses from steam electric power generation are projected to increase substantially and with other uses are likely to cause water shortages in the intermediate to long term in some areas.

Other Effects of Cooling Methods

A number of chemicals are used in conjunction with various cooling systems. In plants using once-through cooling, chlorine which may cause damage to the biota is sometimes added to prevent fouling of the condensers. The possibility of damaging effects is more critical in estuarine and marine areas than in lakes which are in turn more critical than in riverine systems.

The method and timing of passage of water through the plant cooling system is also important. Cooling water intakes for any system should be designed so as to avoid the trapping of mobile species. Improved screening methods should be employed in future cooling water systems to accomplish this. Also, designs should emphasize short travel and exposure times for biota, especially entrained organisms, passing through the cooling system.

It should be noted that high temperature rises are not intrinsically bad if exposure times are short and the effluent dispersion is carried out rapidly within limits acceptable to the key species involved. In addition, the design and construction of discharge outlet structures must take into account their possible adverse effects due to high velocity currents and scour.

Effects of Cooling Towers on the Environment

A great deal of concern is expressed about the possible adverse ecological, climatological, and esthetic impacts of various cooling techniques, particularly cooling towers. Some concern is expressed through purely subjective evaluation, such as the esthetic effects of cooling towers on the landscape. Other aspects can be given more quantitative analysis, particularly with respect to such things as consumptive use and cost.

There is a great divergence of opinion concerning the extent of the effects of cooling towers on local climate through vapor plumes, induced fog, precipitation, and icing. Generally speaking however, there is agreement that these effects are possible. A problem in the past has been inadequate consideration of climatological and meteorological factors in plant location and cooling tower design. For instance, there is a much higher potential for problems if power plants are located in areas of high smog and air pollution potential, especially in small valleys where there is a propensity for stagnation of overhead air masses.

In the operation of cooling towers, a number of chemicals are used as corrective measures to prevent or reduce such things as wood deterioration, biological growths, corrosion, scaling, and general fouling. These chemicals, as well as others naturally occurring in water, tend to become concentrated by evaporation and are released to water bodies as what is termed blowdown. Blowdown becomes an industrial waste which must then be subject to control under water quality standards.

One direct effect of cooling towers is that their use lowers the efficiency of the power plant thereby requiring either (1) more fuel input at the plant or (2) more plants overall, to make up for the lost energy production. This then becomes a necessary part of evaluating (1) future energy demands and (2) the total environmental impact of this, or any, type of cooling facility.

STEAM ELECTRIC POWER PLANT SITING ALTERNATIVES

A broad range of possibilities in siting future power plants exists. Many of these are presently under study by industry and government agencies. It is recognized that while water resources are only one consideration in siting, the potential impacts are of major concern, and therefore, should be given detailed evaluation.

Simulation and Predictive Modeling

An important aspect of site evaluation is the need to assess the probable impact of various cooling system alternatives on the desired use of the water resource. Simulation and predictive modeling techniques, while still in the embryonic stages of development, show significant promise for describing thermal life support system interaction and the processes governing heat movement in aquatic systems.

Continued development and refinement of these techniques are progressing with a view to their leading to (1) better definition of possible problems before plants are designed and constructed, and (2) better heat discharge designs to mitigate adverse environmental impacts.

Pre- and Postoperational Studies

Another important aspect in assessing the possible impact of heat on the water environment is the opportunity to learn from actual experience gained in power plant operation. This can be most meaningful if carried out through a regularized process of pre- and postoperational investigations and monitoring studies correlated with the predictive modeling mentioned earlier. This would allow the comparison of actual effects with those predicted and thereby allow a refinement of modeling techniques which would in turn lead to better plant designs.

These studies should be conducted so as to provide needed information on the effects in situ of imposing a power plant discharge on the water environment; this would then aid in assessing effects of future sites and in evaluating a range of condenser cooling alternatives.

Siting Alternatives

Although there is currently a great deal of attention on new siting alternatives, a large number of future power plants using present technology will undoubtedly and justifiably locate on inland rivers, natural lakes, man-made impoundments, and along the coast.

A number of possibilities exist for previously untried combinations of separately proven concepts as well as completely new concepts. A number of studies and proposals have been made for siting nuclear plants in ocean, coastal, and estuarine areas. In discussing ocean, coastal, and estuarine siting, it is common to refer to:

(1) ocean siting as siting on or in the ocean

(2) coastal siting as land-based plants on smooth or relatively smooth shorelines with ocean outfalls, and

(3) estuarine siting as siting in areas where there is a mixing of fresh and salt water such as embayments with coastal drainage and/or upland river inflow and unrestricted river entrances.

The new concept include floating power plants, power plants constructed on man-made islands on the continental shelf, as well as underwater plants resting on the ocean floor. In addition to placing less of a demand on fresh water sources, i.e., consumptive loss is no problem, the heated discharges may provide some positive benefit.

The upwelling of heated water may push the nutrient-laden lower layers to the surface and through this overturning enhance the fishery resources. Siting alternatives of this type deserve special attention for the intermediate and long term.

If placed far enough in the ocean, heated discharges will be dispersed by the ocean currents and diffused into the large volumes of water available. In order to minimize environmental impact, however, it is important that they be discharged beyond the surf zone and also not in areas where littoral drift and/or near shore currents will return the heated waters to the shoreline, with the exceptions of those areas where net benefits are known to exist.

In addition, a number of plants have been and will likely continue to be proposed for estuarine areas. It is believed that a much larger range of problems presents itself in this case. Estuarine areas, besides predictable esthetic disadvantages, because of their delicate ecological balances, are generally much more susceptible to damage through man's activities than many other areas, and should therefore be viewed with caution.

Presently untried is the possible combination of pumped-storage with nuclear power generation either above or below ground. Variations of this could possibly use waste treatment plant effluent as a water source with ultimate cooling water disposal through deep ocean outfalls or through pump-back as in pumped-storage developments.

POTENTIAL AND/OR DEVELOPING TECHNOLOGIES

The quantities, locations, and effects of waste heat on the water resource are very dependent on the following technologies: generation technology, which indicates through its energy conversion efficiency the amount of heat to be discharged; cooling technology, which determines how the heat is to be transferred from the condenser to the atmosphere; and transmission technology, which has an impact on where the generating facility is located thereby locating the source of the heat discharge. Potential beneficial and multiple use technologies are also important from the viewpoint of their possible effect on reducing waste heat discharge and/or combining waste heat discharges with other inputs in fully designed systems. Each of these are addressed in the above order.

Generation

Nuclear power is expected to play an increasing role in meeting the electrical energy needs of the United States with 50 to 75% of total capacity by 2000. The greatest portion of this is expected to be generated by light water nuclear reactors, 30 to 40% of total capacity, with high temperature gas-cooled reactors comprising 10 to 20% of total capacity.

Nuclear breeders may well be providing 10 to 15% of total capacity by year 2000. Breeder technology, while having a higher efficiency than the current average efficiency of steam electric generation, will approximate an efficiency of only 38 to 42%. While it does have other advantages such as no chemical or particulate air pollution, more efficient use of uranium and thorium, and conservation of fossil fuel resources, it will provide little relief from the overall waste heat problem.

Under present planning assumptions, the thermal energy of a breeder reactor is to be converted to electrical energy using conventional (steam turbine) Rankine cycle technology. However, efficiencies using this cycle are limited because the limitations of metals involved preclude obtaining steam temperatures much in excess of 1050°F, the equivalent of an available fossil fuel steam plant.

A promising technology which represents a potential revolutionary change in future power systems is the fuel cell. It is a device which converts the chemical energy in gaseous fuels directly to electrical energy, thereby avoiding some of the heat engine efficiency limitations imposed by the second law of thermodynamics. It is not a large central station generation method and therefore lends itself to unique possibilities such as small-scale units for apartment houses, shopping centers, and small residential areas.

Magnetohydrodynamics (MHD) is another promising technology. Instead of a solid conductor rotating in a magnetic field, a jet of high-temperature high-velocity ionized gas is forced through a magnetic field. By placing electrodes in this hot gas stream, direct current at relatively high voltages can be obtained. MHD can be used as a topping cycle for conventional steam cycles or by itself; either way the efficiency is 40 to 50% higher than with present fossil fuel plants and would therefore have a significant effect in reducing heat rejected to the atmosphere. A number of serious developmental problems remain, however, and therefore MHD is not seen as being a large contributor to generating capacity in the near term.

For the longer term, more efficient generation technologies appear to hold promise. The most promising is fusion power which appears to go the farthest in reducing waste heat discharges. Although it is difficult to assess the quantity of waste heat discharge, it is expected to range from 5 to 25% of the total heat generated, a significant reduction from present technology. Even with an expanded R&D program, however, it is not expected to be a major contributor to central station technology before 2010.

Cooling

The potential for reducing the heat discharge to water bodies through the development of new technologies in cooling systems at present generally lies in improvements to systems already developed or under development. No revolutionary designs appear on the horizon. The substantial improvements possible in the planning and design of known systems include such things as improved intake and discharge systems, more efficient and more environmentally compatible cooling towers, and better dispersion and diffusion methods thereby allowing greater use of once-through cooling.

The development of efficient economical large-sized dry cooling towers would have the most significant impact since with dry towers there would be a negligible consumptive loss and no heat discharge to water; for cooling these towers depend on the convective transfer of heat to the air as it passes through a fin-tube heat exchanger, rather than on evaporation.

Unfortunately, present dry tower designs are quite costly to operate and maintain, especially since they reduce the plant's average annual energy output by 6 to 8% (and require the construction of an additional 12 to 16% installed capacity).

Design breakthroughs in lowering the cost of dry tower cooling, if achievable, could be of great significance. It is important to point out, however, that there may be basic physical problems such as the much greater inefficiency of air as a cooling medium and the possible environmental impacts from discharging dry heat from large point sources.

Transmission

New transmission technologies have the potential of allowing much greater flexibility in siting power plants, thereby providing a broader range of more economical alternative site locations from which to choose. For instance, these technologies may allow the more economical location of power plants long distances from load centers in order to take advantage of more abundant water supplies. They would also help make beneficial and multiple use technology more economical where such uses involved resource opportunities far removed from major metropolitan and power load centers.

Until recently, it has been uneconomical to transmit power over long distances because of excessive lines losses which in turn increase power generation requirements. Recent developments, however, have now made possible the aboveground long distance transmission of very high voltage direct current with only moderate losses. Since esthetic factors have become quite important in transmission line construction, there is increasing pressure to place transmission lines underground.

Under present technology, however, underground or underwater transmission is quite costly, including a greater line loss potential, with estimates ranging from ten to forty times the cost of overhead lines.

In addition to underground transmission, the development of such things as superconductivity, SF_6 gas-cooled transmission, and the combined cryogenic electrical and liquified gas concept would be significant breakthroughs in allowing greater flexibility in the consideration of siting alternatives.

Beneficial Use Technology

In general, recycling or reuse of waste products present a number of interesting and desirable possibilities for the future. In this vein, a wide range of applications for beneficial uses of waste heat from steam electric power plants have been identified and investigated, and will be discussed in forthcoming chapters.

This concept has been the subject of a number of recent conferences and contract studies. Possibilities have been suggested for the use of heated water in aquaculture, mariculture, agriculture, airport defogging and deicing, deicing of shipping lanes, extending the recreation season, acceleration of sewage treatment, space heating and air conditioning, and for industrial processes.

Viewed from the perspective of trying to solve the waste heat problem, beneficial uses must either (1) reduce adverse effects on the water environment or (2) provide an economic gain or payout to help allay the cost of necessary special cooling systems. Unfortunately, there are a number of constraints which make it difficult for either of these criteria to be met.

With respect to the first point, it should be emphasized that use of waste heat does not reduce the amount of heat to be rejected. All energy, whether wasted or used is eventually rejected to the biosphere.

On the second point there are a number of problems. One is the relatively low quality heat available in the cooling water discharge. Normal maximum temperatures range from 90° to 100°F, with a high of 120°F. Most potential users (one exception being sprinkler irrigation) require a much higher temperature for efficient utilization systems.

A second problem is the large quantities of low grade heat produced. For instance, it has been estimated that to use the waste heat from a 1,000 MW plant for food processing would require 70 to 80 ordinary-sized food processing plants, and if used for greenhouse agriculture could heat 4.4 square miles of greenhouse.

Thirdly, there is frequently quite a difference in the timing of maximum heat discharge and of the need for waste heat. Furthermore, even though waste heat may be available when needed (if the plant is operating), a large part of the time the heat may not be needed; auxiliary cooling would, therefore, still be required. An example would be where heat is needed only infrequently as with airport defogging and deicing.

Fourthly, while certain technologies have a potential for use in connection with completely engineered new systems such as space heating and air conditioning, new cities development and multipurpose agro-industrial complexes, many are beyond engineering and economic practicality when trying to apply them to existing systems.

Fifthly, a potential problem with any beneficial use technology is the possibility of a forced shutdown of the power plant thereby eliminating the waste heat supply.

A related development with considerable potential is the total energy concept. This involves an on-site electrical energy generation system which captures the waste heat produced and converts it into steam or hot water and uses this by-product for heating, air conditioning (through such things as absorption chillers), and domestic hot water. This noncentral station generating technology would, like the fuel cell, provide a revolutionary alternative to present trends of larger central station generation and point source waste heat discharge.

Multiple Use Technology

A number of possibilities also exist for incorporating the location and operation of steam electric generating plants with the operations of industrial plants, the production of agricultural commodities, mineral and mining processes, desalting as well as commercial ventures.

Obviously, developments of this type can seldom be add-on processes. For greatest efficiency they must be completely engineered at the outset. Developments of this type have a number of possible advantages, the most significant to this study being more efficient use of the energy resources and use of the waste heat produced from the electrical energy generation process.

Thermodynamics of the Electric Power Generation Cycle

Mention has been made of the inherent inefficiencies in power generation. Below, a discussion is provided which outlines the reasons for these inefficiencies. This description is based on a study prepared by Dr. R.W. Zeller for the Federal Water Pollution Control Administration and published as PB 215,412 January 1970.

THERMAL ELECTRIC POWER GENERATION AS A WASTE HEAT SOURCE

All major thermal electric power plants in the United States operate on the Rankine cycle and all follow the general pattern schematized in Figure 3.1. As can be seen in this figure, the primary difference between fossil-fueled and nuclear-fueled power plants is in the heat source for steam generation.

Typical, modern fossil-fueled boilers provide steam at 3,000 psi and 1000°F. Because of reactor safety requirements, the current generation of nuclear-fueled reactors of the boiling water or pressurized water types produce steam at about 600°F and about 1,000 or 2,000 pounds per square inch, respectively.

Figure 3.2 shows the basic Rankine cycle. Inputs and outputs are labeled according to the functional parts of the power plant diagram in Figure 3.1 as follows: **3** to **4** is the work input provided by the feed water pump to the boiler; **4** to **1** is the thermal input provided by the boiler; **1** to **2** is the conversion of thermal energy to mechan-

FIGURE 3.1: SCHEMATIC OF BASIC RANKINE POWER CYCLE

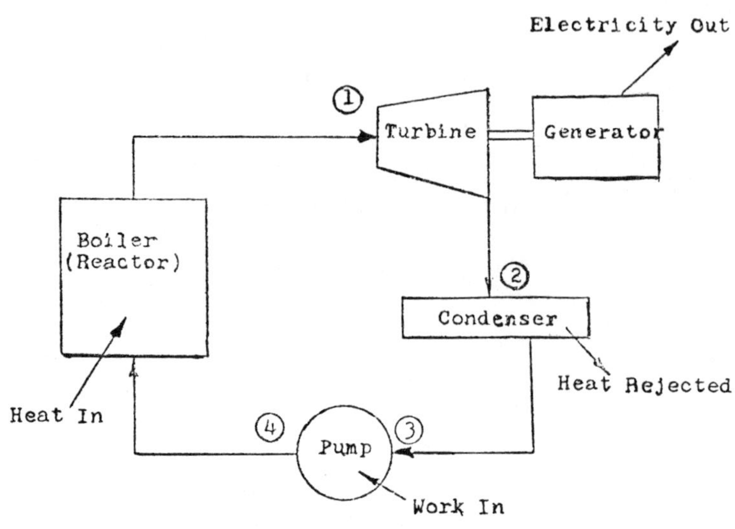

FIGURE 3.2: TEMPERATURE ENTROPY DIAGRAMS OF RANKINE CYCLES

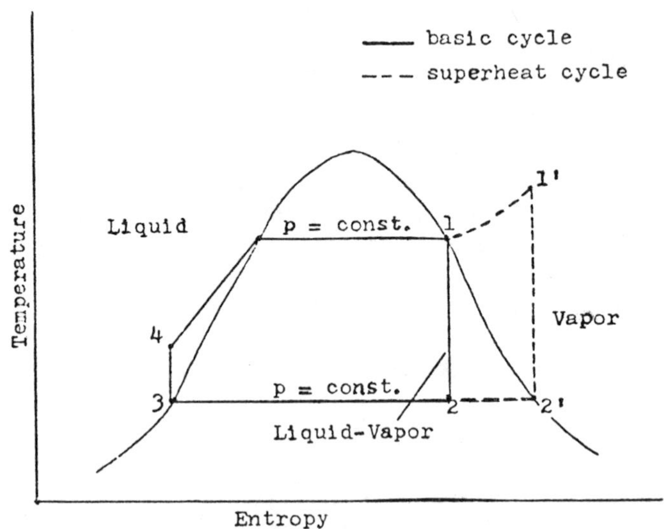

Source: Report PB 215,412; January 1970

ical/electrical energy by the turbine-generator units; and **2** to **3** is the heat rejection incurred by condensing spent steam to water for recycling.

Thermal efficiency of the basic Rankine cycle is calculated as the quotient of the work output divided by the thermal input as follows:

$$E_{th} = \frac{W_{turbine} - W_{pump}}{Q_{in}} = \frac{(h_1 - h_2) - (h_4 - h_3)}{(h_1 - h_4)}$$

where h is the enthalpy, or heat content, of the respective points on the Rankine cycle. The maximum efficiency of this cycle, corresponding to present fossil-fueled power plant design, is about 42% and presumes superheating the steam as shown in Figure 3.1 by the dashed line extension of points **1** and **2**.

Maximum thermal efficiency of the present generation nuclear power plants is only about 33%. An alternative relationship for calculating thermal efficiency is to divide the thermal equivalent of electrical energy output by the thermal input as follows:

$$E_{th} = \frac{\text{electricity output}}{\text{thermal input}} = \frac{3{,}413 \text{ Btu/kwh} \times 100}{3{,}413 \text{ Btu/kwh} + \text{waste heat (Btu/kwh)}}$$

The denominator of this efficiency equation is called the heat rate of a plant and represents the average amount of heat required to produce 1 kwh of electricity. Not all of the waste heat is discharged to the receiving stream, however.

Some of the waste heat in fossil-fueled plants, about 10%, is discharged with the stack emissions; an additional 5% is wasted within the plant as radiation and other losses. Inplant losses for nuclear power plants are estimated at 5%. On this basis, then, waste heat discharged with the condenser cooling water can be calculated as follows.

Fossil-Fueled Plant –

heat to cooling water = (0.85 x heat rate - 3,413) Btu/kwh
at 40% efficiency

$$\text{heat rate} = \frac{3{,}413}{0.40} = 8{,}533 \text{ Btu/kwh}$$

heat to cooling water = (0.85 x 8,533) - 3,413 = 3,800 Btu/kwh

Nuclear-Fueled Plant –

heat to cooling water = (0.95 x heat rate - 3,413) Btu/kwh
at 33% efficiency

$$\text{heat rate} = \frac{3{,}413}{0.33} = 10{,}340 \text{ Btu/kwh}$$

heat to cooling water = (0.95 x 10,340) - 3,413 = 6,400 Btu/kwh

The difference in waste heat rejection to the cooling water between fossil-fueled and nuclear-fueled plants is obviously significant; 65 to 70% greater for the nuclear plants. The importance of this difference is driven home by the prediction that nuclear-fueled power plants will provide two-thirds of the thermal electric energy requirements by the year 2,000.

Another obvious conclusion from the above numbers is that thermal electric power generation is extremely inefficient, resulting in huge quantities of wasted energy. Improvements in fossil-fueled power generation efficiency are limited by available steam conditions in the Rankine cycle as described above. Modern fossil-fueled plants are approaching the practicable limit on thermal efficiency.

Nuclear-fueled plants can and will be more efficient with third and fourth generation power reactors, using gas or liquid metal for primary coolant instead of water. Maximum efficiency of the high temperature gas/metal power reactors is still limited, however, by the Rankine cycle at about 42%.

There are several alternatives to the conventional Rankine cycle, which are in various stages of development and use, as already mentioned. Included among the alternatives are electric power generation by magnetohydrodynamics (MHD) and fuel cells. None are projected to be of major importance in the foreseeable future.

Gas turbines (jet engines) are being installed in power-peaking units and do not reject waste heat to water cooling systems. These units are relatively inefficient, however, and are not expected to replace conventional thermal electric units for base power generation.

THERMAL POLLUTION EFFECTS

It is important at this point to dispel any notion that general temperature increases in the aquatic environment can ever be described as thermal enrichment. Not that temperature increases under certain circumstances cannot be considered beneficial, they are. The danger in using a term like thermal enrichment lies in the hazardous conclusion that only excessive temperature increases are bad.

While we argue over what constitutes an excessive temperature increase, disaster may strike in the form of fish kills, unwanted algal blooms, or unacceptable water supplies for specific municipal and industrial uses. With this in mind, the following paragraphs include examples of specific physical, chemical, and biological responses to thermal pollution.

Gas solubilities are inversely proportional to water temperature; the saturation level for dissolved oxygen (DO) is reduced 50% with a water temperature increase from 32° to 90°F; almost 0.1 mg/l/1°F temperature rise. Dissolved nitrogen behaves similarly to small increases in water temperature and with lethal effects to fish under conditions of dissolved nitrogen supersaturation. Water temperatures influence algal populations directly according to the following temperature preferences:

diatoms (Chrysophyta)	59° - 77°F
greens (Chlorophyta)	77° - 95°F
blue-greens (Cyanophyta)	96° - 104°F

The blue-greens are particularly unacceptable as a group; consequently a shift in population dominance to blue-greens is considered adverse.

Most saprobic bacteria (responsible for decomposition of organic materials) and parasitic bacteria are below their optimum temperature ranges at normal water temperatures in the United States. Parasitic bacteria, particularly, prefer temperatures from 86° to 104°F. Consequently, water temperature increases favoring these undesirable bacterial forms must be considered adverse.

Temperature effects on fish and shellfish are numerous, too numerous to discuss in detail here, and can be categorized according to life stage and geographical distribution of individual species.

WASTE HEAT TREATMENT

For thermal electric power plants located on inland fresh waters, there are only two practicable alternatives for waste heat treatment, cooling ponds and cooling towers. As implied above, direct discharge of condenser cooling water to receiving streams with inadequate dilution should not be considered as an acceptable alternative. In fact, in many locations, the cooling water cycle should be closed with no residual waste heat discharged to the receiving stream.

Cooling ponds can be a relatively low cost, effective, multipurpose mode for waste heat treatment. Generally speaking, cooling ponds can be specifically designed impoundments for this purpose or result from effective utilization of existing impoundments.

In either case, they can serve other functions, including recreation, sports fishing, and flow regulation for downstream users. In terms of overall impact on the environment, cooling ponds are definitely recommended.

Cooling ponds specifically designed for this function should be channelized to maintain flow through circulation, thereby taking advantage of the exponential relationship of heat dissipation to water surface temperature. Required surface area for these flow through cooling ponds can be estimated as follows:

$$A = \frac{Q}{k} \times \ln \frac{\Delta T_o}{\Delta T_d} \text{ acres}$$

where Q = cooling water flow, AF/day;
k = heat transfer coefficient (2.0 ft/day, for example);
ΔT_o = temperature rise across the power plant, °F;
ΔT_d = temperature difference between pond discharge and plant intake, °F

For a 1,000 MW power plant (Q = 2,000 AF/day, ΔT_o = 30°F and an acceptable residual temperature (ΔT_d) of 3°F, the calculated surface area is 2,300 acres. This is close to a commonly used yardstick estimate based on two acres per MW, or 2,000 acres total for a 1,000 MW plant.

For comparison, the required surface area for a completely mixed pond (uniform temperatures throughout) can be calculated as follows:

$$A = \frac{Q}{k} \times \frac{\Delta T_o}{\Delta T_m} - 1$$

where ΔT_m = the difference between pond temperature and plant intake temperature, °F

Calculated surface area for the same 1,000 MW plant would be 9,000 acres, consequently, the recommendation is to design flow through circulation insofar as possible.

For most cooling ponds, the circulation patterns will be somewhere between flow through and completely mixed. Theoretically, then, the average cooling pond should be larger than 2,000 acres for a 1,000 MW plant. In fact, however, the design engineers may compensate to some extent for pond circulation pattern handicaps by concentrating power plant discharges at the water surface.

The heated surface layer takes additional advantage of the exponential temperature heat dissipation relationship. Induced stratification offers a second advantage of permitting cooler water withdrawals at power plant intakes located on the pond bottom.

Where adequate land is unavailable for cooling ponds, wet-type cooling towers are an acceptable, moderate cost alternative for waste heat treatment. The functional parts of wet cooling towers used in large power plant installations include: inlet water distribution system, a packing layer to increase water-air contact surface area, inlet air louvres, drift (carryover of water droplets with tower vapor) eliminator vanes, cooled water basin, and air movement equipment.

Mechanical draft towers regulate air flow by means of large fans. Natural draft towers (commonly hyperbolically shaped) induce air flow by density differences between the air-water vapor mixture inside the tower shell and ambient air.

Both mechanical and natural draft towers, with numerous variations, can be designed to effectively treat power plant cooling water.

Agricultural and Aquacultural Uses of Waste Heat

The information given in the following two sections on agricultural and aquacultural uses of heat rejected from large central station electric generating plants was obtained from a study performed by M.M. Yarosh, B.L. Nichols, E.A. Hirst, J.W. Michel, and W.C. Yee of the Oak Ridge National Laboratory and published as ORNL-4797 in July 1972.

AGRICULTURAL USES

In contrast to many urban and industrial heat applications which require heat at temperatures higher than is normally wasted from electric generating or other steam process plants, several agricultural uses (e.g., spray irrigation, soil heating, and environmental control of animal shelters and greenhouses) offer a way to use the thermal discharges without reducing electrical energy production.

For example, steam or hot water at 300°F is needed for the various urban energy requirements currently under consideration. If steam from back-pressure turbines is extracted at 300°F rather than at 100°F, the gross turbine cycle efficiency is decreased from 47 to 30% for a modern fossil-fueled plant, and from 33 to 18% for a light-water reactor nuclear plant.

Power plants with cooling towers are normally designed so that the temperature of condenser effluent is between 80° and 120°F; plants with once-through cooling operate near these temperatures much of the year but may discharge water as low as 55° or 60°F in the winter. In many locations these temperatures are high enough to provide satisfactory thermal environments for many plants and animals. Thus, agricultural operations which can be located close to the power station may truly be considered potential waste heat users.

Spray irrigation and soil heating can be used to lengthen the growing season and provide frost protection in certain regions. Maintaining animal shelters at the proper temperatures can increase growth rates and feed efficiencies; this is particularly important for the smaller animals, such as poultry and swine. Greenhouse production of both flowers and vegetables is critically dependent on artificial heating and cooling, and the use of waste heat from condenser cooling water can significantly reduce fuel costs to greenhouse operators.

In spite of these obvious benefits from the use of waste heat for agricultural applications, several potential obstacles exist. Most important, the current level of agricultural production is such that only a small fraction of the waste heat discharged from power plants can be used profitably, and the projected growth patterns suggest that this picture will not improve in the future.

In addition, the use of waste heat is strongly dependent on geography, climate, and season. The size of the greenhouse, poultry, or swine operation required to use a reasonable fraction of the waste heat generated by a typical power plant is an order of magnitude larger than any installations in the U.S. today. However, several foreign countries do have greenhouse operations that could use all the exhaust heat from a several hundred megawatt electric plant.

Future greenhouse operations in the U.S. may be able to use the waste heat from power plants to replace the dependence on gas and other fuels which are in short supply. Such large operations, however, may introduce new problems in management, disease control, and waste disposal. Also for some operations, temperature control is not a controlling cost, and the lure of cheap (or even free) heat may not be sufficient to attract agricultural operations to power plant sites.

Finally, certain probems may arise in coupling the power plant operation and the agricultural operation. The utility may be reluctant to have a second party on its cooling system or to obligate itself to supply the warm water from a nuclear plant where concern on the liability for radioactive contamination may be a problem.

Nevertheless, agricultural uses of waste heat are sufficiently attractive, under certain conditions, to warrant serious consideration. While these uses will not solve the "thermal pollution" problem, they can, in particular locations, reduce the impact of thermal effluents on the local ecology, conserve energy resources (reducing demand for fossil fuel heat), and save money for both the electric utility and the agricultural operator. The potential and problems associated with the use of waste heat for open-field agriculture, greenhouses, and animal shelters are discussed in the following sections of this chapter.

Open-Field Agriculture

Throughout the history of agriculture, man has generally been at the mercy of nature and has adapted to or accepted the vagaries of the weather or climate in his particular area. Control of temperatures in agricultural activities was initiated only recently and is still limited primarily to greenhouse horticulture and poultry operations. The importance of environmental control has long been recognized and studied, but the high cost of heat and equipment has limited its application to a few high-income crops.

Considerable study has been devoted to the effects of irrigation water (and soil) temperature on crop production and, of course, to the technical aspects of design and operating techniques intended to minimize the temperature changes. It is, however, important to recognize that much additional work is required as evidenced by the following statement, "Knowledge of how root temperature affects plant growth is woefully incomplete, partly because of ignorance about root function."

The idea of using waste heat from power stations for agricultural purposes was suggested at least as early as 1957, but it is only recently that several investigators have begun to study and evaluate the potential benefits, costs, and problems. As a consequence, little information in the literature is specifically directed toward the productive use of waste heat in field agriculture. Nevertheless the use of warm water for field irrigation through subsoil or sprinkler application techniques represents potential applications for the waste heat in discharge water.

Incentives for Use of Waste Heat in Open-Field Agriculture: Both plants and animals respond to specific environmental conditions that are conducive to optimum growth. Although these conditions vary considerably among species and at different growth stages, it is seldom that optimum values are maintained in nature. One of the important variables influencing plant growth is the temperature of both soil and air, and although this discussion deals primarily with temperature, it should be recognized that many other critical environmental factors interact with temperature and may require simultaneous adjustment as the temperature approaches an optimum level. These factors include soil moisture, air humidity, plant nutrition, and soil-air and atmospheric-air composition.

It is known that within certain temperature ranges, biologic activity essentially doubles with each temperature increase of 10°C, but that too low or too high temperatures are lethal to plants.

Potential benefits of temperature control to open-field agriculture include the following:

(1) Prevention of damage caused by temperature extremes,
(2) Extension of the growing season,
(3) Promotion of growth,
(4) Improvement of crop quality,
(5) Control of some diseases and pests.

Prevention of Temperature Extremes — Perhaps the most obvious and the most easily adapted use of waste heat in the form of warm water is in frost protection, particularly to tree crops. Frost protection by sprinkler application depends on the "heat of fusion", that is, the release of heat by water as it freezes. A critical factor in this technique is the proper management of water application to preclude limb or stalk breakage from ice formation, which can cause losses that exceed the losses caused by frost. The use of warm water in the spray system can alleviate this problem.

Warm water can also be used to cool plants during the hot, dry summer months when atmospheric humidities are low. In such a situation it has been demonstrated that warm water applied through a sprinkler system attains ambient temperature by the time it reaches the soil surface. Heat loss results from evaporation, cooling both the plants and the water supplied through the sprinkler.

The use of warm water for the purpose of controlling crop damage due to extreme temperatures, while of vital agricultural importance, generally is required only a few hours during a few days of a year. Thus, this application requires a highly reliable heat source which is used at a very low load factor, and would present problems in capital cost amortization, unless the warm water can be distributed through an irrigation system which would have been needed anyway.

Expansion of the Growing Season and Promotion of Growth — Temperature is one of the most important factors governing the germination of seeds. Germination, emergence, and early growth of plants are intimately related to soil temperature. The effects of weather are probably more critical during germination and early seedling development than during any other stage of vegetation growth. Unfavorable soil temperatures at seeding time often produce a poor stand and consequently a reduced yield.

Retarded growth of young seedlings may not only further reduce yield but also adversely affect the quality of crop produced. Favorable temperatures at the seedling growth stage may enhance growth sufficiently to provide the possibility for producing two or more crops per year, and thus greatly increase farm income. Also, achieving earlier crop maturity can give a large marketing advantage, particularly for high-value crops.

As indicated above, basic knowledge of the relationship of soil temperature to plant growth is limited. Some literature is available on this subject, although much of it is related to plant growth in the noneconomic sense, that is, rate of net photosynthesis, total dry matter accumulation, time of (or percent) seed emergence, root volume, etc.

There is, however, some literature that discusses yields. In one experiment, rice (grain) yields were increased from 32 to 55% by increasing the root temperature from 18° to 30°C. Corn yields (silage) increased by 68% when the soil temperature was increased from 12° to 27°C, while for potatoes (tubers) the yield increased by 47% for a soil temperature change from 12° to 20°C, but then decreased by about 40% when the temperature was further increased to 27°C.

Table 1 summarizes some recent data based on field experiments. While all yields obtained in these experiments were depressed by water shortage, the growth in the heated corn plots was particularly restricted by insufficient irrigation.

Based on the observed rate of yield increase and past experience of corn silage production potential in this region, yields of 13 to 15 tons/acre of dry matter are considered attainable. Soil warming also added to the quality of the product, and the nitrogen content in the silage was increased.

TABLE 1: EFFECT OF WARMING SOIL ABOVE ITS NATURAL TEMPERATURES

Crop	Yield (tons/acre)		Yield increase (%)
	Unheated	Heated	
Corn			
Silage	5.5	8.0	45
Grain	3.2	4.3	34
Tomatoes	32.1	48.3	50
Soybeans, silage	2.25	3.74	66
Bush beans			
First planting	6.44	7.80	21
Second planting	3.30	5.70	73
Total	9.74	13.50	39

Two crops of bush beans were grown in succession on the same area, but the second crop on the unheated area did not mature. The beans harvested on the unheated plots were extremely small and could not have been sold commercially.

On the other hand, the second crop on the heated plots was of the same quality as the first. Based on these observations, bean yields of 12 to 15 tons/acre, or more than twice the unheated yields, are considered feasible. Even higher yields probably can be obtained by using optimum practices and high-density planting.

Improved Crop Quality — Crop quality is believed to be, in part, a function of the overall plant growth cycle, and therefore control of the environment over the plant's lifetime should improve the final product. Little is known, however, about the specific effect of elevated root temperature in commercial crop production. Crop quality may not necessarily be increased, since plant production of certain materials is not assumed to be a single-valued function of root temperature over the whole range of plant growth temperatures.

Control Disease and Pests — Cool soils tend to encourage certain diseases, particularly in cotton, and coolness also adversely affects the quality of the fiber. The use of water heated above the temperature usually considered in the "waste heat" range, has been proposed for soil sterilization. For example, the golden nematode (and its eggs) are killed by 5 minutes exposure to water at 125°F (49°C).

With lower temperature exposures (105° to 110°F) it may take 20 to 60 minutes to be lethal.

Potential Problem Areas: Most of the incentives for controlling soil temperature and the resulting benefits are of considerable significance. A basic problem exists in the economic risks due to the undemonstrated techniques and crop yields on large farms over extended operating times; that is, the overall economics have not been established, particularly with regard to the high initial investment where irrigation would not normally be needed.

Also long-term testing may reveal problems in soil management or plant disease and pest control, although there is presently no indication of such problems.

A significant question appears to exist in the breadth of application of waste heat from power stations for soil temperature management or irrigation. Most power stations are located relatively near population centers in areas of adequate rainfall where irrigation is only supplemental.

On the other hand, many power stations are located in the higher latitudes where the soil warming feature could be advantageously used. Generally, the western part of the U.S. is deficient in rainfall, and particularly in parts of the Northwest both soil warming and irrigation appear to offer good potential for the use of waste heat from power plants.

There are also problems in continuously utilizing the entire flow of warm water [a 1000-MW(e) power station would continuously discharge 500,000 to 700,000 gpm] on nearby farms. However, even if the entire flow could not be distributed continuously, the careful selection of sites in regions of arid agriculture could benefit large farming areas. Since irrigation results in some consumption of water by evaporation and transpiration, the use of condenser water on previously unirrigated land might be objectionable.

Such a use and its benefits would have to be weighed against the alternate choice of wet cooling towers and their water makeup and blowdown requirements, and the comparison of pumping and piping cost would have to be determined for soil irrigation cooling. In the western states, the availability of water would have to be determined, and legal restrictions would have to be defined.

There may also be questions raised on the direct agricultural use of cooling water from nuclear plants from the standpoint of potential radioactive contamination. Special precautions may be necessary to prevent this problem from occurring.

Problems of power plant operation, refueling, shutdowns, and effect on the power conversion cycle efficiency have not been adequately analyzed, and additional costs or areas in which research is needed may be revealed.

Conclusions: While agricultural uses for power station waste heat appear to be beneficial, especially for arid areas where water quality standards prohibit the return of heated water to streams, the current state-of-the-art for open-field use is quite limited.

Most of the required research and development areas have been identified, and initial results are encouraging. However, areas in which work may be needed are: (1) induced power-station problems such as increased pumping costs, siting restrictions, etc.; (2) the economics of the total operation, including marketing and projected price structure of agricultural products; (3) overall ecological effects; (4) additional legal restrictions resulting from the combination of power production and irrigation.

Since several of the research and development projects mentioned are being sponsored by power utility companies, it would be expected that these problem areas are receiving attention, but as yet they have not been discussed in the literature.

Greenhouses

The utilization of waste heat for greenhouse operation has been suggested in several studies. Since an exclusion area is required for nuclear power plants, it has been suggested that greenhouses might be constructed on this idle land adjacent to nuclear plants to use the waste heat from the power plants and under certain circumstances might conceivably replace cooling towers which would otherwise be required.

In areas where the heating costs amount to from 10 to 30% of the operating cost (or $2,000 to $11,000 per acre) for greenhouse production of vegetables, the potential reductions in cost of heating provide a considerable incentive to develop large greenhouse operations in conjunction with power plants. This arrangement would allow the use of otherwise wasted resources (heat and land) without reducing the efficiency of the power plant. The increasing difficulties in obtaining natural gas (a primary fuel for heating in greenhouses) provide additional incentive for looking to the use of waste heat for heating and cooling of greenhouses.

Crop yields can be greatly improved through the utilization of heat, in controlled-environment glass or plastic houses, providing an added incentive for the use of waste heat. However, because the amount of heat available from large power stations is so great, it is not likely that the construction of greenhouses would be practical for the dissipation of all the waste heat produced by the power plants being constructed today.

Incentives for Using Waste Heat in Greenhouses: The use of greenhouses for the culture of vegetables enables larger crops and crop yields (up to ten times the open-field output) to be realized with small amounts of land area.

In addition, the ability to culture crops the year round allows more uniform productivity and permits the matching of crop harvest with periods of high demand and high price. Providing plants with the optimum temperature can reduce the time required to produce a crop and can greatly improve the yield per plant. Optimum temperatures vary with the species.

Vegetables cultured at warm daytime temperatures of 80° to 100°F and nighttime temperatures of 75° to 80°F include squash, watermelon, cantaloupe, and cucumbers; those cultured at daytime temperatures of 75° to 85°F and 60° to 65°F nighttime temperatures are tomatoes, peppers, okra, eggplant, and onions. At low daytime temperatures of 70° to 80°F and nighttime temperatures of 50° to 60°F, spinach, radishes, cabbage, broccoli, carrots, beans, beets, lettuce, and cauliflower are cultured.

Although many crops can be grown in greenhouses, the differences in value tend to encourage intensive production of only a few species. These include tomatoes, cucumbers, and lettuce. Growth curves for these three plants are shown in Figure 4.1. The costs of producing vegetables in greenhouses vary with location, but the largest items in the operating costs are always labor and fuel. With heating costs reduced, profits may be increased, or more money may be available to pay for increased costs of labor or materials. The use of waste heat from steam-electric power plants therefore appears promising as a source of low-cost heat for use in greenhouses, especially if the plants have cooling towers with wintertime operating temperatures of 60°F or higher.

FIGURE 4.1: IDEALIZED GROWTH CURVES

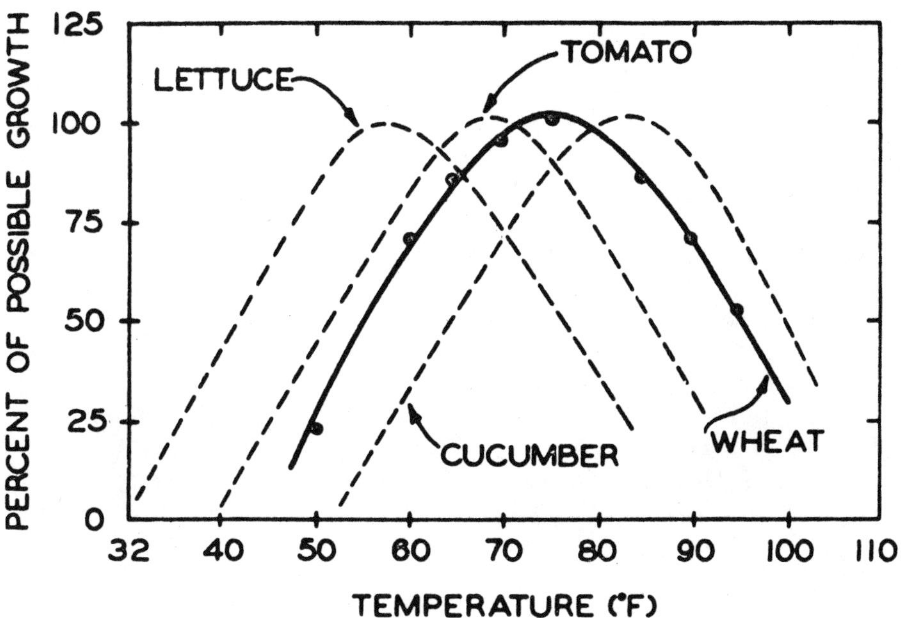

Source: PB 210 747, April 1972

The extensive use of greenhouses in or near areas of high population density would permit supplying food to nearby markets at seasonably favorable periods. Labor is considered the greatest single problem in the industry, and it may limit extensive use of greenhouses in some areas.

Current Greenhouse Practices and Designs: Current greenhouse operations employ either glass or plastic-covered houses. Recent interest in plastic-covered houses results from their low capital and construction costs and their tax advantage. Developments in plastics and the use of twin-layer plastics for greenhouses have reduced the heating costs by reducing the heat losses through the roof (double-layer plastic houses require only 5/8 the amount of heat required for glass houses). Greenhouses have been heated with hot water for many years by using radiators or finned tubes. In addition, water is often used for cooling with evaporative pads and fans. Studies have shown that warm water in the pad and fan system, in conjunction with finned tubes, can be used for both summer cooling and winter heating.

The best documented studies of the use of waste heat in greenhouses involve the joint efforts of the University of Arizona and the University of Sonora at Puerto Penasco, Sonara, Mexico. At Sonora the waste heat of diesel engine, generator sets is used in a desalting plant and the growing areas of the controlled-environment greenhouses.

Environmental control is provided in the University of Arizona experiment by use of a direct-contact heat exchanger in which air is forced through packed columns into which seawater is sprayed at the rate of 120 gpm. Variation in flow rate is used to regulate the temperature.

When warmer temperatures are required, the 94°F blowdown water of the desalting plant may be used instead of seawater. In this system the humidity remains at nearly 100%, and the air temperature is close to that of the water passing over the packed-column heat exchangers. Approximately 20,000 cfm of moist air is circulated for ventilation and temperature control in the greenhouses. Warm water from power plants and other industrial processes could be used for such agriculture.

A preliminary feasibility study of the use of warm water for heating and cooling greenhouses in the Denver area was carried out by Oak Ridge National Laboratory. The study showed that the cooling tower planned for the 330-MW(e) Fort St. Vrain nuclear plant of the Colorado Public Service Company could be replaced with low-cost (relative to cooling tower capital cost) evaporative heat exchangers located in the greenhouses.

For the design wet bulb temperature in the Denver area (65°F), calculations indicated that the greenhouses could be cooled to at least 75°F in the summer by evaporating 92°F water (available from the turbine condenser) with once-through air, the heat being discharged to the outside. By recirculating the greenhouse air through the evaporative pads during the winter, the air temperature could be maintained above 65°F with a 0°F outside temperature. In either mode of operation, heat dissipation is constant and "full load".

In the ORNL design, water is circulated through evaporative pads as shown in Figures 4.2 and 4.3. Air passing through the pads is evaporatively cooled during periods of high ambient temperature and heated during periods of low ambient temperature. To maintain the humidity at levels under 80%, finned-tube heat exchangers are placed downstream of the evaporative pads so that dry heat could be added to reduce the air humidity.

Discussions with plant physiologists and horticulturists suggest that plants are more prone to fungus and disease at humidities above 85%; hence the need for humidity reduction. However, the University of Arizona experiments have been with nearly saturated air. The water returning to the condenser approaches the wet bulb temperature of the air during operation, and the water heats or cools the air moving through the pad, depending on water temperature and entering air temperature conditions.

FIGURE 4.2: PAD ASSEMBLY

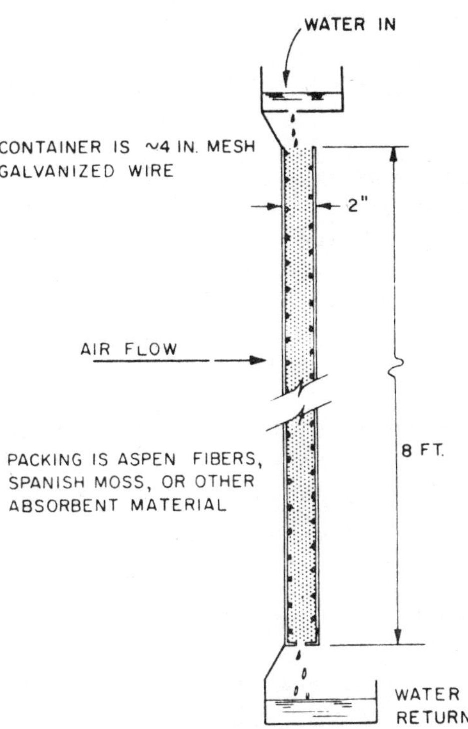

Source: ORNL-4797, July 1972

FIGURE 4.3: TYPICAL GREENHOUSE WITH AIR AND WATER FLOW SYSTEM

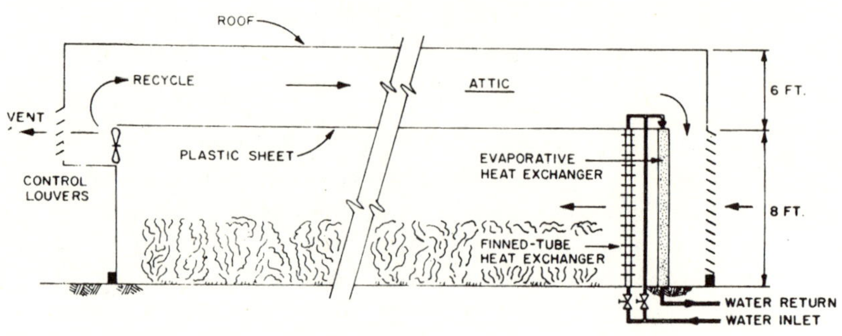

Source: ORNL-4797, July 1972

Figure 4.3 shows the greenhouse arrangement. Except for the plastic sheet used for the attic to permit air recycling, the arrangement is fairly typical of large greenhouse units that use evaporative pads for summer cooling.

During the summer, air enters the greenhouse through the pads and exhausts at the opposite end. As outside temperatures drop, the discharge louvers close and force the air to recycle through the attic and subsequently through the evaporative pads. During cold nights the relative humidity of the air leaving the pads is nearly 100%, and the finned-tube heat exchanger is used to heat the air to reduce the humidity in the greenhouse to 80 to 90%.

Table 2 gives the calculated air and water conditions for several summer operating cases, with evaporative pads replacing the cooling tower of the Fort St. Vrain plant. It was assumed that hot water from the condenser would be piped directly to the greenhouses. The flow rates expressed are for each 50 x 100 ft greenhouse. In all cases the range of the temperature is 22°F, the same as for the Fort St. Vrain plant, which as design temperatures of 80° to 102°F.

Table 3 gives similar data for winter operating conditions. Data for wind and sky conditions are also given in the table, and the relationship between the air, water, and roof temperatures is shown. The heat available from the Fort St. Vrain plant would be enough for 250 to 300 acres of greenhouses.

TABLE 2: GREENHOUSE CONDITIONS FOR SUMMER OPERATION

Case	Ambient conditions		Air flow rate (lb/hr)	Water flow rate (lb/hr)	Range of conditions in greenhouse		Range of water temperature (°F)
	Dry bulb temperature (°F)	Relative humidity (%)			Temperature (°F)	Relative humidity (%)	
1[a]	95	16	306,000	88,200	76–86	80–67	67–89
2[b]	50	73	306,000	88,200	~58	~95	51–73
3[b,c]	50	73	153,000	88,200	~67	~100	57–79
4[d]	95	16	306,000	44,100	71–81	85–71	64–86
5[e]	50	73	306,000	44,100	~53	~90	48–70
6[f]	50	73	153,000	44,100	~57	~100	50–72

*Data are for each 50 × 100 ft greenhouse.

[a]Summer conditions for Denver (64°F wet bulb) and 500 MW of waste heat dumped to 100 acres of greenhouses.

[b]Moisture in air assumed to remain same as for day conditions, but dry bulb temperature dropped to 50°F.

[c]Air flow rate reduced by one-half.

[d]Conditions same as in case 1, except that 200 acres of greenhouses were assumed and the water flow rate was reduced by one-half.

[e]Similar to case 2, with 200 acres of greenhouses and water flow rate reduced by one-half.

[f]Similar to case 3, with 200 acres of greenhouses and water flow rate reduced by one-half.

TABLE 3: GREENHOUSE CONDITIONS FOR WINTER OPERATION

Wind velocity: 15 mph
Effective sky temperature: 100°F
Greenhouse area: 200 acres

Outside air temperature (°F)	Water flow rate (lb/hr)	Air flow rate (lb/hr)		Air temperature (°F)		Range of water temperature (°F)	Mean roof temperature (°F)
		Recycle	Vent	Over plants	Through attic		
−30	44,100	153,000	0	72	72–65	66–88	1
−15	44,100	153,000	0	76	76–69	71–93	15
0	44,100	153,000	0	80	80–74	75–97	26.5
0	44,100	148,400	4,600	72	72–65	66–88	21
0	44,100	141,400	11,600	63	63–56	56–78	15
0[a]	26,500	153,000	0	56	56–50	51–73	12
0[a]	26,500	148,400	4,600	51	51–44	45–67	8.5

*Data are for each 50 × 100 ft greenhouse.

[a]Emergency conditions: reactor shut down and an emergency heater being used to supply heat at the rate of 1.5 MW/acre.

During the summer, water returning to the power plant from the pads would be cooler than would normally be delivered by the existing cooling tower and therefore would increase the efficiency of the plant during hot weather.

Summer cooling conditions are favorable for the Denver area, because of the design wet bulb temperature of 65°F. It should be noted that the water temperature from the pad approaches the wet bulb temperature of the air within 3° to 5°F. In areas having a high wet bulb temperature, the cooling effectiveness would be less than in areas with a low wet bulb temperature.

Current Development Programs: While there are no large greenhouse operations in this country using low-level heat from power plants, experimental work is being carried out which could lead to large-scale use in the future. As mentioned earlier, the University of Arizona and the University of Sonora are using heat from diesel generators to provide heat for greenhouses at Puerto Penasco, Sonora, Mexico, where crops have been grown at near 100% relative humidity.

The success led to a request from the Sheikdom of Abu Dhabi for construction of a 5 acre facility on the island of Sa'Diyat in Abu Dhabi, and this system is now in operation.

One of the unique features is the ability of the facility to conserve water through collection of the condensate which occurs on the plastic roof. It is reported that each 4600-ft^2 greenhouse will yield up to 1,500 gallons of water per day during periods when the exterior temperatures are low enough to result in condensation on the inside of the plastic roof.

Since this is distilled water which can be recovered and used for makeup water, during the winter there would be a potential recovery of water amounting to $\cong$14.2 thousand gallons per acre of greenhouse per day. This water could also go to supplying the approximately ten thousand gallons per acre per day irrigation needs of the crops being raised, and to provide high-quality makeup water for the power plant cooling system.

Although the ORNL feasibility study described earlier indicates that several advantages exist for using greenhouses to cool reactor condenser water, no plans exist to indicate that greenhouses will be built in the U.S. to use a sizable portion of power plant waste heat. However, regardless of whether the power plant is cooled significantly, even a fraction of the heat should be an attraction to a greenhouse operator because of low heat costs. Furthermore, there are many industrial processes and cooling towers wasting heat which could be used in greenhouses at a small expense to the grower.

Recently, an experiment was started at Oak Ridge to determine the actual operating performance of a pad and fan system for use in heating and cooling greenhouses. Waste heat in the water from a building air-conditioning system is being used for temperature control in a small plastic greenhouse. Preliminary results thus far have revealed small differences between the theoretical calculations used in the feasibility study and the pad performance, but additional work is required to prove details of the system. The Tennessee Valley Authority is planning a pilot test of the Oak Ridge System of heating and cooling in a joint TVA-ORNL program.

Each of the systems mentioned involves the flow of water from the power plant to the greenhouse, where the water is cooled and sent back to the power plant or discharged to surface waters. The systems available for blending and controlling the water to maintain certain temperatures require conventional engineering. The use of greenhouses in series or parallel with cooling towers, cooling ponds, or other systems could afford increased flexibility for waste heat use.

Although work is being conducted throughout the United States on design of greenhouses, greenhouse equipment, and on growing methods, little work having direct applicability to the utilization of waste or low-temperature heat from power plants is currently under way. Sufficient information exists to design a greenhouse system to use waste heat. However, the integrated performance of large complexes may require on-site demonstration facilities before many questions can be answered.

Evaluation and Summary of Use of Waste Heat for Greenhouses: A principal advantage of using waste heat from power plants for greenhouses is that it does not require modification of the plant and does not reduce power cycle efficiency. The publication is one of several options available for the dissipation of waste heat produced during the production of electric power or other industrial processes.

However, if all of the commercial greenhouses existing in the U.S. today used waste heat, they would consume only a few percent of the heat being wasted from existing power plants. Since the growth rate for power plants exceeds the present growth rate for greenhouses, it is questionable whether more than 1 to 5% of the total waste heat could be utilized by greenhouses, and therefore the primary incentives must be economic rather than a solution to the thermal discharge problem.

Although a large amount of water would be consumed in greenhouse operation, water losses would be less than for cooling towers if the water condensing on the greenhouse surface were collected and returned. As described earlier, during recirculation in winter, most of the water could be recovered from condensation in the attic.

In the studies at Puerto Penasco, the closed environment itself greatly improved the yeilds of a wide variety of crops even though relative humidity was nearly 100%. Most successful varieties were those developed in hot humid areas. Tomato varieties such as Floradel, N-65, and Tropic did well, while varieties such as Michigan-Ohio, Wolverine 119, and Tuckcross-0 did not.

Whether operation at 100% humidity is possible, in colder cloudy areas of the country and with other varieties, remains to be seen. High humidity at night can result in the collection of water on the leaves of plants. This may result in growth of fungi and the spread of bacteria which are likely to be detrimental to the plant.

During winter the greenhouse operator must depend on a reliable supply of heat. At power sites with multiple units the reliability of the heat supply should be high. During scheduled or unscheduled outages of a unit, heated water would be available from alternate operating units.

Base-load nuclear plants with high reliability seem aptly suited to the greenhouse requirements. Nuclear stations are equipped with a fossil-fired heater of 100 MW or more, which would provide additional reliability. In some cases a separate emergency heating unit would have to be provided.

The use of low-temperature heat therefore represents a potential way of significantly reducing operating expenses and increasing profit for the grower. A very large greenhouse operation could in turn reduce the capital investment and operating expense of the power plant operator by providing a substitute heat rejection system and a market for previously wasted heated water. Thus gains might be realized by both parties in a greenhouse operation large enough to use a sizable fraction of the waste heat from a power station, but the investment required would be large. For example, glass houses which used one-fourth of the waste heat from a 100-MW(e) power plant would require a capital investment of approximately $25 million and occupy about 250 acres. Although no such large installations are expected for many years in the United States, it is reported that single operations of 250 acres exist in Hungary.

Little work has been done to date on the evaluation of the market for greenhouse-produced crops at the scale necessary for using such massive quantities of waste heat. Most of the existing data are extrapolated from small-scale operations of 5 acres or less.

There are many unanswered questions concerning the use of waste heat from power plants. Chemicals such as chromates used for water treatment in the cooling water system might affect the plants in a greenhouse.

Similarly, the pollen from the greenhouse could possibly affect the cooling system. The determination of whether such effects will occur requires experimental studies. In the case of nuclear plants the real and imagined hazards of radioactivity must be considered, and public acceptance of products produced in such greenhouse complexes would have to be analyzed. Potential sources of activity in the cooling water would have to be considered and measuring devices installed to continuously monitor the water for radioactivity.

The most difficult questions to resolve appear to be those of institutional arrangements necessary for the financing and operating of such an enterprise in conjunction with the operating of a power plant. The organization and training of the greenhouse operating teams, agreements with the utility on shutdown schedules, provision for auxiliary heat supply, and protection of the power plant coolants from loss or fouling are several of the important problems. If risk insurance is common to greenhouse operation, the degree to which it might be affected by coupling to a power plant for heat would have to be determined.

All of these questions point to the necessity of conducting research or studies to resolve uncertainties which now exist. Although engineering questions can be resolved fairly easily, these and the biological and economic questions require demonstration projects with corps in a greenhouse facility.

Marketing data, legal restrictions, economic incentives, insurance, and other questions need extensive probing before the full potential can be ascertained. Labor is considered the number one problem. Problems of providing the large skilled staff necessary for a successful operation must be investigated and solved.

Conclusions: Adequate engineering information is available to allow the design and operation of a heating and cooling system for greenhouses utilizing waste heat from steam-electric power plants. Prospects and incentives exist for the coupling of greenhouse vegetable operation with electric power production. The principal uncertainties are in the marketing problems related to high production rates, institutional arrangements for implementing such a program, and the problem of public acceptance of the product.

Presently there is need for detailed examination of the operation of a large-scale greenhouse complex in order to resolve these questions.

Animal Shelters

The feed efficiency (pounds gain/pound feed) and growth rate of some farm animals are strongly dependent on environmental temperature. Proper temperature control can decrease feed consumption and increase productivity. This is particularly important for small animals (with a large surface area to volume ratio) such as poultry and swine, and considerably less important for cows. Because the production of other farm animals (e.g., sheep, goats) is small, only poultry and swine production will be discussed here.

Poultry Operations: During the past several decades, broiler production has become concentrated in fewer, but larger farms. A typical operation today might produce 40,000 to 100,000 birds annually. Broiler production has grown spectacularly in recent years, from 6 billion pounds in 1960 to 11 billion pounds in 1970, an increase of 80%.

Table 4 lists the eight leading states in broiler production. Production is heavily concentrated in the Southeast; almost 60% are grown in Georgia, Arkansas, Alabama, North Carolina, and Mississippi. The reasons for this geographic concentration are probably related to low labor costs and a warm climate.

Typical costs to the farmer of producing broilers are tabulated in Table 5. This table shows the importance of maintaining high feed efficiency. Feed accounts for 62% of the total cost of raising broilers.

TABLE 4: U.S. BROILER PRODUCTION (1969)

	10^6 lb	Percent of total
Georgia	1,548	15.4
Arkansas	1,410	14.1
Alabama	1,235	12.3
North Carolina	1,038	10.3
Mississippi	774	7.7
Maryland	680	6.8
Texas	597	5.9
Delaware	521	5.2
Rest of U.S.	2,243	22.3
Total	10,046	100.0

TABLE 5: BROILER PRODUCTION COSTS

	Cents/lb sold	Percent of total cost
Fixed costs: depreciation, interest, taxes, insurance, maintenance	1.0	5.7
Chicks	3.1	17.6
Feed	11.0	62.5
Labor	1.4	8.0
Fuel	0.6	3.4
Miscellaneous	0.5	2.8
Total cost	17.6 cents[a]	100.0

[a]These costs are based on 1962 data for the New England area, which may not be representative of current national costs.

Poultry Physiology — The value of a controlled environment for broilers has been recognized for some time. Numerous experiments have been conducted which show that feed efficiency and growth rate can be improved by properly adjusting the temperature, humidity, and ventilation within poultry shelters. Experimental investigations have demonstrated the importance of temperature control for young chicks. Figure 4.4 shows the effect on growth rate of changing air temperatures for chicks between the ages of 9 and 18 days. Air temperature was always 87°F on the 9th day.

Other studies have shown that maximum growth rate and feed efficiency occur between 60° and 70°F for broilers four weeks of age and older. These results are summarized in Figure 4.5. Figure 4.5 shows that increasing the temperature from 40° to 60°F increases broiler growth rate by 14% and feed efficiency by 11%. This suggests that proper temperature control can reduce feed costs (improved feed efficiency) and reduce per unit labor and capital costs (higher growth rates).

FIGURE 4.4: EFFECT OF TEMPERATURE ON THE GROWTH OF CHICKS

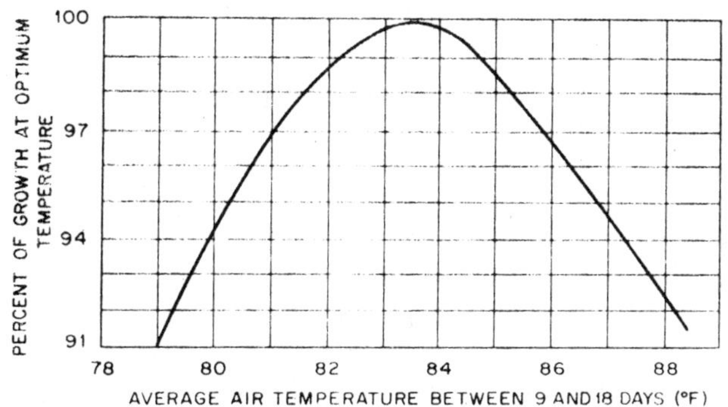

Source: ORNL-4797, July 1972

FIGURE 4.5: FEED CONVERSION AND WEIGHT GAIN FOR 4 to 8 WEEK OLD BROILERS

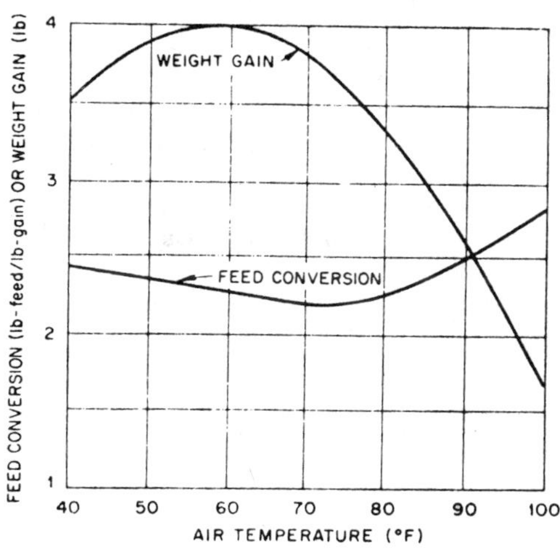

Source: ORNL-4797, July 1972

This shows that broilers grow best when the temperature is maintained within the appropriate temperature range, assuming the humidity is maintained between 50 and 70%. Adequate ventilation is required to remove moisture and odors, and to provide a uniform temperature distribution and adequate oxygen. Ventilation rates should be about 1 cfm/lb in winter and 2 cfm/lb in summer.

Current Shelter Engineering Practices — A well-insulated house can cut fuel costs by a factor of 4 compared with an uninsulated house. Research demonstrated that energy requirements per broiler ranged from 20,000 Btu in the summer to almost 60,000 Btu in the winter with an uninsulated house for a full eight week period, that is, from birth to market. These figures were reduced to 5,000 and 13,000 Btu for an insulated house.

Current design recommendations for broilers include proper ventilation, adequate insulation, and the use of brooders for young chicks. Installed brooder capacity ranges form 15 to 30 Btu/hr-bird, depending on insulation and geographic and climatic conditions. Summer cooling is usually accomplished with increased ventilation rates, although evaporative cooling is used in some locations.

Swine Operations: Hog production has remained fairly constant over the past several years, increasing only slightly from 96 million in 1955 to 102 million in 1970. Hog prices received by farmers have varied erratically over the past 15 years, however, the trend seems to be toward an increase in hog prices.

Table 6 lists the eight leading hog-producing states. Hog production is very concentrated; 70% of the hogs produced come from the eight midwestern states listed in Table 6. Production is concentrated in these states because of the availability of inexpensive feed, primarily midwestern corn.

Currently, a large hog operation produces about 5,000 pigs/year. In the past, hogs have been grown in two annual shifts, a spring and a fall crop. Recently, the trend has been to year-round growing to make better use of the farrowing houses.

Economic data concerning hog production is quite scanty. Table 7 shows that feed accounts for 65% of the costs in raising hogs. Fuel accounts for about 4% of this cost.

TABLE 6: U.S. HOG PRODUCTION

	10^3 head on farms 12/1/70	Percent of total
Iowa	16,322	24.2
Illinois	7,468	11.0
Indiana	5,129	7.6
Missouri	5,120	7.6
Minnesota	3,692	5.5
Nebraska	3,691	5.5
Ohio	2,838	4.2
Kansas	2,202	3.3
Rest of U.S.	21,078	31.1
Total	67,540	100.0

TABLE 7: APPROXIMATE DISTRIBUTION OF HOG PRODUCTION COSTS

	Dollars/hog sold	Percent of total cost
Buildings, equipment	$ 3.60	18
Feed	13.00	65
Labor	1.60	8
Fuel	0.80	4
Veterinary medicine, miscellaneous	1.00	5
Total cost	$20.00	100

Swine Physiology — In a study on the influence of ambient air temperature on the growth rate of swine, the optimum temperature of hogs has been shown to vary from 73°F for 100 lb pigs to 65°F for 250 lb pigs. The growth rate drops off sharply on either side of the optimum temperature and even becomes negative at high temperatures.

For example, above 95°F larger pigs begin to lose weight. This occurs because of depressed appetites and increased respiration. The growth rate of a 200 pound pig decreases by 33% if the temperature is more than 13°F higher or lower than the optimum (69°F).

Current Shelter Engineering Practices — Traditionally, environmental control in hog houses has been limited to ventilation and insulation. Several USDA publications on swine shelters were issued in the late 1950's and make no mention of supplemental heating or cooling for noninfant swine. Infrared brooders were recommended for warming baby pigs, and shades and wallows were suggested for cooling pigs in summer.

More recently, supplemental heating has been recommended for all swine in winter to improve weight gain and feed efficiency. Heater capacities of 2,000 to 5,000 Btu/hr for a sow and litter, and 100 to 500 Btu/hr for a finishing swine are suggested.

Experiments performed in South Carolina using electrical strip heaters showed an average (based on a year's operation) supplemental heating requirement of 1,400 Btu/hr for a sow and litter confined in the heated pens for 21 days.

Thus, current trends in hog production are toward greater environmental control. Specifically, some form of supplemental heating (under-floor or hot air) is becoming increasingly prevalent for both farrowing and finishing operations. Protection from thermal stress in the summer is normally obtained by high ventilation rates, proper insulation, and drinking water. Neither mechanical air conditioning nor evaporative cooling appears to be widely used.

Potential Benefits of Waste Heat Utilization: It is of interest to compute the fraction of waste heat produced by steam-electric power plants which can profitably be used for temperature control of swine and poultry shelters.

Approximately three billion broilers were grown in the U.S. in 1970. Since broilers are grown year-round, the average energy required to brood all these chicks can be taken as 10,000 Btu/chick for a well-insulated house. Thus approximately 0.3×10^{14} Btu/year are required to brood all the broilers currently grown in the U.S., using current data. With low-cost waste heat available from power plants, heat use might be higher, since the economic penalty of providing optimal thermal conditions would be greatly reduced.

Various estimates were 5.3×10^{15} Btu/year of waste heat were rejected from power generating stations in 1970. Thus, about 1% of the total waste heat generated could be used for raising broilers, under current conditions if all the broilers were raised using waste heat from power stations. In the winter, broiler heating could use almost 2% of the waste heat discharged, but in the summer it could use only 0.5% of this heat.

Estimates that 5,000 Btu/hr are desirable for a sow and litter during a typical Iowa winter. This heat would be used for the equivalent of 50 days at the above rate. Assuming 5 million litters/winter implies that 3×10^{13} Btu would be required for heating all the sows and litters produced in winter. This compares quite well with 0.7×10^{13} British thermal units suggested by data obtained in the much warmer climate of South Carolina.

In addition, approximately 300 Btu/hr are required for finishing pigs. Using the same 50 day full use factor and 50 million pigs per winter gives 1.8×10^{13} Btu required for the finishing operation in winter.

Thus, about 5×10^{13} Btu/winter are required to supply the current winter heating needs of American hog production. This is 1% of the total waste heat generated and about 3% of the winter heat generated. During the summer, very little waste heat would be required, just enough to keep the litters warm at night. The thermal requirements of the hog production are slightly greater than those of broiler production. However, broiler heat requirements are not so concentrated in the winter.

The use of waste heat can reduce fuel bills and increase feed efficiency and growth rate for both hogs and broilers by providing optimal temperature conditions. A pad and fan system, in conjunction with a finned-tube coil (system described later), can provide both winter heating and summer cooling while, at the same time, cooling the condenser water.

The Evaporative Pad and Fan System: The system envisioned for heating and cooling animal shelters involves the use of conventional pad and fan systems with finned-tube coils; see Figure 4.3 and the discussion on greenhouses in this report. Pad and fan systems are currently used in many greenhouses and in some poultry and swine operations for cooling purposes.

The pads (see Figure 4.2) are typically filled with a semipermanent fibrous material. Condenser cooling water flows onto the pads from a trough at the top and drips vertically down along the fibers. Air flows horizontally across the pads and is heated or cooled depending on the ratio of sensible to latent heat transfer. The cooled water is collected at the bottom of the pads and in a closed system would be pumped back to the condensers. Warm water from the condenser may also be pumped through the finned-tube coils, located downstream of the pads. Alternately, the warm water could be run through pipes embedded in the floor of the shelter. The air coming from the pads is heated and dried by the transfer of sensible heat across the fins.

By varying the relative fractions of water pumped through the pads and the coils and the air flow rate, the temperature and humidity of the air entering the animal shelter can be adjusted over wide ranges. This system can be used for both summer cooling and winter heating. The heated (or cooled) air passes through the house and out the other end through exhaust fans. Automatically controlled louvers would permit recirculation under conditions of extreme cold.

With this system the environment within the animal shelter can be maintained near the optimum. Simultaneously, the power plant condenser water is cooled, approaching the ambient wet-bulb temperature. Thus, the animal shelter serves as a horizontal cooling tower.

Problems: A significant obstacle to the use of waste heat for animal shelters is insufficient knowledge. Further studies are needed to determine the technical and economic feasibility and desirability of such a system.

Using current figures, broiler houses and swine shelters could use about 2% of the total waste heat generated at steam-electric power plants if all present animals were raised using such heat. The generation of electricity has been doubling every ten years for the past several decades and will probably continue to do so for some time. The growth rate in swine production is considerably lower, only a few percent per decade.

Broiler production has increased rapidly, about 80% during the past decade. However, this growth rate is also slower than the growth in electrical generation. Thus, it appears that in the future, animal shelters will require an even smaller fraction of the waste heat generated, assuming current trends continue.

Geographic concentration is another factor which may inhibit the use of waste heat for animal shelters. Hog production is very concentrated in the Midwest, and broiler production is concentrated in the Southeast. Power plants in these areas may be able to couple their operations with agricultural enterprises, but throughout most of the country, broiler and swine production are so low that they will be unable to use more than a small fraction of the power plant waste heat. However, it is possible that the lure of cheap (even free) heat may induce broiler and swine production to shift geographically.

For example, New York produces only 0.1% of American broilers but probably consumes 5 to 10% of the total production. If the use of waste heat can lower production costs sufficiently, New York may be able to grow its own broilers.

In order to minimize pumping and piping costs, the broiler and swine operations would have to be located adjacent to the power plant (within the exclusion area for a nuclear plant). The waste heat from a 1000-MW(e) plant is sufficient to brood almost one billion broilers a year or farrow and finish about 10 million hogs a year.

As indicated earlier, a typical broiler operation currently produces about 50,000 birds annually, and a large hog operation produces about 5,000 pigs/year. Thus, current operations are two or three orders of magnitude smaller than would be required to use 10% of the waste heat from a modern power plant.

Several problems may arise with large operations such as disease, odor, and waste disposal, and these have not yet been resolved. In particular, waste disposal may be a major problem with hog operations. Current legislation and regulations require improved waste treatment, and the resulting economic penalty may inhibit the development of larger operations. However, future technological developments may eliminate this problem.

Similarly, hog operations require a considerable amount of land. Estimates have shown that a 1,000 hog operation requires about 30 acres. This includes hog housing, feed storage, and waste disposal facilities for a controlled-environment operation. Linear extrapolation indicates that 30,000 acres would be required to produce a million hogs/year [enough hogs to use 10% of the waste heat from a typical 1000-MW(e) plant].

The capital costs of the pad and fan and finned-tube coil system plus the pumps and piping are higher than the costs of conventional brooders and space heaters. These additional capital costs must be compared with the reduction in operating costs due to the use of waste heat. In many locations (e.g., the South) the additional capital expenses may not be justified.

The demands for heat in animal shelters are quite seasonal, considerably higher in the winter than in the summer. Yet it is during the warm summer months that thermal pollution problems are most severe. In the summer the animal shelter would serve as a horizontal cooling tower, with little advantage to the farmer. In fact, the high humidities and temperatures associated with this operation may be detrimental in certain regions of the country where the wet-bulb depression is small.

Variations in electrical generation may seriously hamper the use of waste heat for heating animal shelters. If the power plant shuts down for a long time during a period when heat is required, alternate means must be provided for warming the chicks or pigs. The cost of installing a backup heating system must be compared with the savings from the use of waste heat. This would not be a problem at multi-unit power plants.

Biocides and other toxic substances are usually added to condenser cooling water to prevent the growth of algae within the condenser tubes and accessory piping. Carry-over from the evaporative pads may be harmful to poultry and swine.

During the winter, when air is being recirculated within the animal shelters, high dust levels may accumulate on the pads and in the cooling water. This dust buildup may block airflow, reducing heat transfer, and may also change the chemical quality of the cooling water sufficiently to aggravate corrosion problems.

The condenser cooling water circulated through the pads in the animal shelter is cooled largely by evaporation. This represents a consumptive use of water, amounting to about 2% of the total flow rate. In arid regions this water loss may be unacceptable. However, the water losses are no higher than they would be with an evaporative cooling tower.

Climatic variability is another factor which may inhibit the use of waste heat in animal shelters. Only certain regions of the country have climatic conditions suitable for the use of waste heat. The Midwest is a good candidate for waste heat applications, because the winters are cold and the summers are cool, with reasonable wet-bulb depressions. The Southeast, on the other hand, has warmer winters and a very small wet-bulb depression. So heat utilization will probably be minimal in the Southeast.

Concern about radioactivity may make people reluctant to buy pork and broilers grown in a reactor exclusion area. This problem can probably be overcome with a suitable public education program.

Conclusions: The use of waste heat for environmental control of animal shelters has the potential for reducing costs and minimizing environmental impacts at certain locations. Potential fuel savings are about $8 million/year for the industry, and the potential reduction in feed costs are in the same range.

However, various problems exist which might inhibit such uses. Studies are needed to determine the technical, economic, and environmental desirability of such a system. Research is needed in several areas to better define the problems and potential associated with waste heat utilization in animal shelters.

Technical questions concerning the actual performance of pad, fan, and finned-tube systems remain. Preliminary work at ORNL suggests that this system can provide adequate environmental control in many parts of the country, but applications to commercial operations must be demonstrated.

The problems associated with economics and management have not yet been addressed. Research is needed to answer the following questions: Can feed efficiencies be further increased or are current practices nearly optimal? What are the problems associated with very large broiler and swine operations? Are such large operations economically viable? Are the savings in fuel costs worth the additional capital expenses associated with pumps and piping? How should the capital costs be apportioned between the utility and the farmer? Will cheap heat reduce the riskiness of these farm operations?

If research suggests that such systems for animal culture are feasible and desirable, then a pilot-plant program should be initiated to obtain field data. Ideally, the field trial should include greenhouse, poultry, and swine operations.

In preliminary trials the size of the operation should be kept small, but in later tests these operations should be significantly increased to about 50 acres of greenhouses, 500,000 broilers/year, and 50,000 hogs per year to reveal the problems caused by larger agricultural operations.

Summary

Agricultural operations are capable of using low-temperature (waste) heat from power plants without reducing electrical energy production. While these uses will not solve the thermal pollution problem, they can, in particular locations, reduce the impact of thermal effluents on the local ecology, conserve energy resources, and save money for both the electric utility and the farmer.

Thermal effluents from power plants can be used in open-field agriculture to promote rapid plant growth, improve crop quality, control pests and disease, extend the growing season, and prevent damage due to temperature extremes. Water, used for both irrigation and heating, can be applied through nozzles (spray irrigation) or through a subsurface piping system. With these systems the farm acts as a large, direct-contact heat exchanger for the power plant, while the utility provides irrigation water to the farmer.

Several research projects are under way in the Pacific Northwest to investigate the feasibility and desirability of these systems. Some additional work is being performed in the Southeast.

This use of heat is of importance for only a few days of the year (early spring and late fall). During the remainder of the year, water is needed for irrigation but not for heating. However, most power plants are sited near urban centers where rainfall is sufficient to obviate the need of irrigation. Also, the long-term implications of waste heat applications for soil management and disease and pest control are not yet known.

The use of power plant waste heat for warming and cooling greenhouses can improve crop growth and yield while reducing operating (fuel) costs by as much as $4,000 to $6,000/acre. With approximately 7,000 acres of greenhouse production today, this represents a total potential saving in fuel costs of $28 to $42 million annually on a national basis (10 to 30% of operating costs).

Research at the University of Arizona, University of Sonora, and the Oak Ridge National Laboratory suggests that using waste heat for greenhouse climate control is both feasible and economically attractive. However, no large-scale field operations are currently under way.

Waste heat can be used to provide optimal temperature control in swine and broiler houses. Fuel costs could be reduced by $8 million annually on a national basis. Additional savings in feed costs may result from improved feed efficiency under controlled environmental conditions.

Additional study is required to determine the limitations imposed on agricultural uses by climate, geography, product marketing, waste heat reliability, effects of biocides and corrosion inhibitors in the cooling water, and consumer acceptance of products grown using cooling water from nuclear plants.

It is essential that these problem areas be thoroughly investigated before a commitment is made to large-scale agricultural applications of waste heat.

AQUACULTURAL USES

Aquaculture is an ancient art. It has been practiced for centuries in the Orient, particularly in the tropical and subtropical areas where farmers raised fish in flooded rice fields to provide a protein supplement to their basic grain diet. Yet the practice is also a new technology.

A few fish species have been intensively cultivated in controlled environments, and yields of these species have been enhanced by the degree of management exercised over the operation. In pond culture, for example, with nitrogen and phosphorus fertilization, yields for carp are 100 to 600 lb/acre-year at sites in Israel and Southeast Asia. With supplemental feeding, these yields increased to 1,600 to 2,400 lb/acre-year.

Most impressive of all are the yields in running-water culture with intensive feeding as practiced by the Japanese. Yields of 0.8 to 0.3 million lb/acre-year for carp have been obtained. Catfish culture in ponds under semicontrolled conditions may yield 2,000 lb/acre-year, while yields as high as 2 million lb/acre-year for catfish and trout might be achievable in intensive culture in a flowing stream with a relatively high degree of environmental control.

By contrast, fishing for wild species on the continental shelf by trawling and purse seining may yield only 20 lb/acre-year. Aquaculture is more like farming, whereas fishing is like hunting. While yields from aquaculture cannot be compared with yields from fishing for wild species, the contrast provides an insight into the potential for aquaculture in supplying future fish demands.

The methods described above are all seasonal activities. No attempt is made to maintain the temperature of the culture system in the optimum range for growth. Yet basic data on fish growth indicate the potential benefits of maintaining optimum temperature (Fig. 4.6).

For example, shrimp growth is increased by 80% when water is maintained at 80°F instead of 70°F, and catfish grow three times faster at 83°F than at 76°F. Growth of both aquatic species benefits appreciably more from temperature control than does growth of animals such as broilers, cows, and swine.

FIGURE 4.6: EFFECT OF TEMPERATURE OF GROWTH OR PRODUCTION OF FOOD ANIMALS

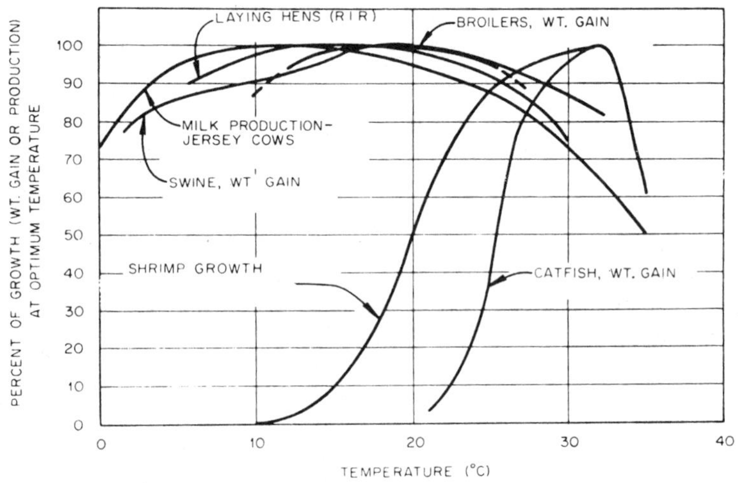

Source: ORNL-4797, July 1972

Heated discharge water from steam power plants represents a large energy source for maintaining the temperature of a culture medium in a range that is optimum for the growth of some aquatic species.

Also, electricity is available for pumping power, permitting greater environmental control over the water system. Thus, thermal aquaculture at power plant sites offers the potential of producing high-quality aquatic foods continuously in some locations, and the possibility of decreasing the present high variability in available supply due to the seasonality of such produce.

Temperature control alone, however, is not sufficient for optimum production of aquatic species. Dissolved oxygen content, biological oxygen demand of the culture system, fish waste control, and nutritional adequacy of the food diet are some of the other important variables that also influence yield.

Methods Used in Fish Culture

Of the 2,500 known fish species, less than 1% of them have been successfully cultured at all, and probably less than 0.5% of them have been intensively cultured as in animal husbandry. The simplest operation is pond culture, in which environmental control is quite limited and variable. Here fish are simply stocked in a body of water.

At low stocking rates (hundreds of pounds per acre), fish can exist on natural food in the water. As the stocking density is increased, nutrient levels of the pond have to be enriched by nitrogen and phosphorus fertilization and addition of supplemental foods.

Aeration may become a necessity to satisfy the increased metabolic oxygen requirements. In addition buildup of fish wastes and oxygen consumption by other aquatic organisms in the system become important factors in overwhelming the system.

In contrast to stationary ponds, dynamic systems of culture can offer a greater degree of environmental control. Fish can be confined in cases (e.g., 30 ft long x 10 ft wide x 5 ft deep) and placed in a large volume of water such as natural lakes and streams, or cooling ponds of channels of coolant water. A greater water flow rate (increased number of volume changes) permits higher stocking density.

Food is fed at regular intervals. Cage culture can result in disease problems when it is carried out in large bodies of water where wild fish populations exist and the fish in culture cannot be isolated.

Flowing water culture is also practiced in multiple channels or raceways, each of which might be 100 ft long x 4 ft wide x 5 ft deep in a commercial operation. Water depth and flow rates can be controlled. The water is utilized more efficiently and productivity (yield/acre) is enhanced. Fish population density can be high.

With flowing water, environmental control is easier than in other systems; that is, dissolved oxygen is distributed more uniformly and biological oxygen demand is lower because fish wastes are flushed away. However, capital costs are about ten times that of the pond or cage culture.

Current Techniques

Catfish is commercially the most widely cultured fish in this country (about 54 million pounds in 1970). This is a warm-water species whose optimum growth temperature is between 80° and 90°F. Seasonal culture is carried out largely in Arkansas, Mississippi, and Louisiana. Farmers usually fill their culture ponds in the spring, stock them with catfish fingerlings, feed the fish during the growing season, harvest them in the late fall, and sell them to processors who market the prepared product.

However, this simple culture method is not without problems. High temperatures and low dissolved oxygen concentrations can result from solar heating of the ponds and pond stratification. Sudden algae blooms can increase biological oxygen demand in the pond system and cause oxygen depletion. Development cost for such ponds in about $400 to $1,200 per acre.

Some of the newer commercial catfish culture projects are more sophisticated in design. Floating cages have been placed in flowing water, or 70° to 75°F groundwater has been pumped continuously into circular tanks. In both types of technology, yields up to 200,000 lb/acre-year or better have been reported.

A successful commercial demonstration of intensive aquaculture is the Thousand Springs Trout Company in Buhl, Idaho. This the largest farm of its kind in the world, supplying 30% of the U.S. market (4 to 5 million pounds in 1969) for rainbow trout.

Yields of 200,000 to 400,000 lb of rainbow trout per acre-year are obtained, and each year shipments of 1.5 million pounds of dressed trout are made to domestic and foreign markets. The year-round culture is made possible by a 250,000 gpm supply of constant-temperature (60°F) springwater that comes from canyon walls. One-fourth of this flow is diverted and distributed into channels where high-density culture is practiced.

At a stocking density of 2 lb/ft^3 of water, the weight of rainbow trout supported in 16 lb/gpm of water flow. Nutritionally balanced pelletized food is fed at regular intervals, and an excellent feed conversion ratio of 1.5 lb of dry food fed per lb of wet fish produced is normally achieved. The commercial operation includes feed formulation and mixture, culture from the egg in the hatchery to growth in flowing water to a uniform marketable size, and processing of the harvested fish to a frozen packaged product.

The uncommonly fresh taste and firm meat of this trout are attributed to the flowing water which flushes away ammonia and nitrogenous wastes. The average wholesale price for the product is in the middle to upper portion of the price range for rainbow trout, $0.85 to $1.15 per pound (1971 price).

The technical success of this enterprise is probably due to the high-quality water at the culture site, coupled with sufficient knowledge about rainbow trout biology to make mass culture possible. There are relatively few sites with such a dependable source of water and only a very few aquatic species whose biological characteristics are known well enough to permit such an intensive operation.

A dependable market for the cultured product is essential to the financial success of the operation. There have been numerous instances of enterprises which were technically successful but failed because of an inadequate marketing arrangement.

Some seawater species have also been cultured on a seasonal basis. Raft culture of oysters and mussels has produced 2,000 to 200,000 pounds of product per acre of water surface along the shorelines of Australia, France, Japan, and the United States. The most favorable sites for these rafts are areas where the nutrient concentrations are enriched by the drainage of rivers and estuaries and where large volumes of moving water surface along to carry natural food supplies to the mobile rafts.

Although no entirely suitable food formula has yet been developed for oysters or mussels, the nutrient content of the water can be further enriched by the addition of nitrogen and phosphorus fertilizers. Yields may be drastically reduced by predator attack (oyster drills and starfish) when the facility is not isolated from the sea. Four years of culture are normally required to produce a marketable oyster.

The Japanese are the foremost fish culturists in the world. Along the bay areas of Japan's Inland Sea, finfish (yellowtail) is cultured in nylon net bags (cage culture) supported by bamboo frames. Oyster culture is an established industry of long standing; rafts are floated in bay areas, and wire strings of oysters are hung from a lattice work on each raft, where the oysters feed by pumping seawater and extracting the available nutrients.

Shrimp culture owes much to the results of a thirty year effort by M. Fujinaga to perfect methods of induced spawning of gravid females and mass hatchery rearing of the larvae forms so that shrimp supplies would not be dependent on the catch of juveniles along the seacoast. Experimental culture of blue crab, abalone, and squid is also in progress.

Several varieties of seaweed (Nori) and algae (Undaria) are cultured by the Japanese for use as a condiment or additive to a variety of foods. Both grow best in seawater that is in the range 50° to 70°F. Monospores cultured in indoor tanks are transferred to nets or strings suspended on bamboo rafts and allowed to grow during the late fall and early winter in shallow estuarine areas. The harvested product is processed into thin dried sheets and sold in packets of 6 inch sheets. In 1967, Nori production was 140,000 tons and Undaria was 67,000 tons.

Heat Utilization in Aquaculture

Thermal aquaculture involves the use of heated effluents (e.g., power plants or thermal springs) to maintain optimal temperatures for growth and produce high yields. Power plant coolant water has only recently been used for aquaculture. A commercial operation, the Long Island Oyster Farms of Northport, Long Island, utilizes the thermal effluent of Long Island Lighting Company for the early stages of oyster culture.

Normal growing periods of four years have been reduced to 2.5 years by selective breeding, spawning, larvae growth, and "seeding" oysters in the hatchery. This avoids reliance on variable natural conditions and permits accelerated growth in the thermal effluent discharge lagoon over a period of about 4 to 6 months, when the water would otherwise be too cold for maximum growth. Oyster culture is completed for market in the cold waters at the eastern end of Long Island Sound.

The product is harvested, processed, and marketed for $15 to $20 per bushel (1971), the upper end of the wholesale price range. 20% of the oysters "set" in the hatchery result in a harvested product.

Catfish have been cultured in cages set into the thermal discharge canal of a fossil-fueled plant of the Texas Electric Service Company at Lake Colorado City, Texas. During the winter of 1969 to 1970, growth rates achieved were equivalent to 200,000 lb/acre-year. This is comparable to the yields of rainbow trout culture in flowing water. The Texas operation is now on a commercial basis.

A pilot research and development project is being conducted by Trans-Tennessee, at the TVA steam plant in Gallatin, Tennessee. Heated discharge water from the plant is circulated through nine of ten concrete channels each 4 ft wide x 4 ft deep and 50 ft in length. Algae formation is minimized by covering the channels and preventing photosynthesis. Presently, studies are being conducted at different stocking densities.

Nutritionally balanced pelletized feed is fed to the catfish in culture. Extrapolated yields of up to 2,000,000 lb/acre-year have been obtained in several of their raceways. The company is planning a 230-channel facility that would supply cultured catfish to the nearby Nashville metropolitan area. The expanded facility would have a capacity for 60,000 lb of dressed catfish per week. With a continuous supply of warm water and a vertically integrated operation like that of the Thousand Springs Trout Company, Trans-Tennessee believes that it can supply the catfish demand of the Nashville area at costs considerably less than pond production costs for catfish.

Large feed production and animal processing companies in the agribusiness industry are considering utilizing waste heat for fish cultivation. Florida Power Corporation of St. Petersburg, Florida, has recently announced a joint five year research effort with Ralston Purina Company to develop a satisfactory technique for culturing shrimp at the utility's Crystal River site.

Armour and United Fruit have conducted a small research effort on shrimp culture in cooperation with the University of Miami at Florida Power and Light Company's Turkey Point facility.

Smaller companies like International Shellfish Enterprises are developing methods for oyster culture in the thermal discharges canal of Pacific Gas and Electric's plant at Humboldt Bay. Marifarms, Inc., of Panama City, Florida, is utilizing the warm water from the local power plant to maintain pond temperatures in winter so that mortality of shrimp in culture is minimized. Experimental lobster culture using warm water is being considered by a few institutions, including a California group (San Diego Gas and Electric Company and Mariculture Research Corporation) and the Department of Sea and Shore Fisheries of the state of Maine.

The Japanese have led the way in demonstrating the benefits of waste heat utilization for aquaculture. Shrimp, eel, yellowtail, sea bream, ayn, and whitefish are being cultured. Culture experiments started at the Sendai Power Plant in 1964. Five other demonstration programs have been established at fossil-fueled power generating stations.

In pond culture at a power plant in Matsuyama, shrimp are cultured in thermal effluents blended with ambient water to maintain constant temperature. Summer growth under culture conditions was 1.2 times the growth of shrimp in natural summer water temperatures, while winter growth was 7 times that of shrimp in ambient temperature water. Survival rates were about 50% in the summer experiment and as low as 30% in the winter experiment.

In flowing water, yellowtail cultured in constant-temperature water from October to June grew to a weight of 1.5 times the weight of fish cultured in natural water. No mortality or parasite problems were encountered.

At the Tokai-Mura Nuclear Power Station near Tokyo, a multispecies $575,000 demonstration program of thermal aquaculture has just been approved by the Japanese government. The five year program is to develop a facility consisting of 35 concrete channels of various sizes to demonstrate flowing water culture. Additional funding for the program is anticipated from the utility companies through the Japanese Atomic Industrial Forum.

The English have had a small development program since 1966 on the culture of flatfish species, plaice and sole, at their nuclear plant in Hunterson, Scotland. The problem of free chlorine toxicity was avoided by the use of a continuous chlorination treatment of coolant water instead of the conventional batch treatment. This resulted in a residual of less than 0.02 ppm Cl_2. The problem can be further reduced if the power plant uses mechanical cleaning techniques.

No radioactivity is allowed to be diluted into the coolant water stream used for aquaculture. Although culture of the flatfish species has been demonstrated, widespread culture has been restricted by low food conversion efficiency and high food costs. Low-value fish is used as feed, and a suitable low-cost, formulated food has not been developed. Flatfish are cultured in flow-through ponds near the shoreline. Since the system is not isolated from the sea, predator attack and disease are of concern.

Feasibility Study of Thermal Aquaculture

A thorough feasibility study (ORNL-4488) of a conceptual design and the market potential for a shrimp culture facility has been performed. The study used the published data on shrimp biology and technology to develop a conceptual design for continuous culture in a flowing stream. A cost estimate, a cost sensitivity analysis, and a market projection for the cultured product were developed.

Sophisticated channel culture was proposed in which juvenile shrimp, cultured from the egg in a hatchery, would be raised in a series of pens of increasing surface area within a channel until the shrimp reached marketable size.

With year-round cultivation at optimum temperature, a shrimp yield of 20,000 lb/acre-year was projected. This would be four times the seasonal yields (lb/acre-year) that have been reported for Japanese shrimp culture in flowing water.

This is based on two crops per year instead of one and the doubling of weight density of shrimp in culture from 55 to 110 g/ft^2 of bottom area (shrimp are bottom dwellers). Weight densities up to 200 g/ft^2 have been reported for the culture of bait shrimp in aerated tanks.

A detailed cost estimate was made for this integrated conceptual design and included feed preparation, shrimp culture from eggs and larvae in the hatchery to growth to a harvestable size in channels of flowing water, and processing to the frozen product.

For the assumptions made, the calculations showed a yield of ten million pounds per year of shrimp which at 1970 market levels would have a wholesale value of $1.00/lb. Production costs were estimated to be about 80 cents per pound.

The production cost was found to be most sensitive to feed conversion ratio and least sensitive to labor considerations; capital costs for site improvement were intermediate. Low-cost, nutritionally balanced feed is important to the economics of shrimp culture, because it constitutes more than 60% of the total operating cost. To date, no feed has been successfully tested for the mass culture of shrimp, although food formulation tests programs are currently under way both in the United States and in Japan.

In this country, formulated feed has been developed only for the mass culture of rainbow trout. This same feed, however, has been used for the mass culture of other fish.

In Japan, shrimp in culture are fed low-value fish which give a food conversion of 10 lb of feed to 1 lb of flesh. It is economically feasible to do this, because retail prices for live cultured shrimp command a higher price in Japan than in the United States.

Market Estimates for Cultured Fish and Seafood

The potential of thermal aquaculture is related not only to technical feasibility but also to markets for the products. There is little statistical data at present to indicate the extent of demand for cultured aquatic foods in this country.

In Japan, fish is a prime source of protein, and the per capita fish consumption in 1967 was 120 lb/year, an order to magnitude above that in the United States. Aquaculture in Japan represents a significant tonnage and monetary value in the fisheries industry.

In 1967, the total catch was 15.6 billion lb, with a value of nearly \$2 billion. Aquaculture products totaled 940 million pounds and were valued at nearly \$300 million, about 6% of the total catch and 15% of the total value. Certain cultured products can command luxury prices in Japan.

As mentioned earlier, the yellowtail fish of the tuna family has been cultured extensively in Japan. In 1963, 60% of the Osaka market for yellowtail was furnished by aquaculture. By 1965, production reached 36 million pounds, but further production increases were threatened by a lack of natural supply of small fry. By 1968, artificial propagation was successfully developed so that future demands for the fry could be met.

Present difficulties in providing a constant supply of fish have presented serious problems to the seafood industry in the United States, and the scarcity of certain seafoods is given as the most serious problem by the U.S. seafood industry.

On a world basis, seafood consumption represents the fastest growing food area, but world sustained yields from natural sources will be limiting for many species within the next few decades. Culturing of fish and other seafoods would reduce this problem.

In the United States, aquaculture is in its infancy. Statistical data show that during the past decade total edible fishery products have risen from 4.3 to 6.2 billion pounds. The domestic catch, however, has remained approximately constant at 2.0 to 2.5 billion pounds, while the imported supply has increased from 1.8 to about 3.7 billion pounds. Less than 1% of the total supply is furnished by fish culture.

However, statistical data provide an incentive for considering the culture of high-value fish species. For example, in the period 1950 to 1970, shrimp per capita consumption rose 160% from 0.8 to 2.0 lb, while total consumption of all seafoods remained relatively constant at about 10 to 12 pounds. Consumption of meat, poultry, and fish combined rose by 40% in this same period.

Although the domestic catch of shrimp is the largest in the world, imports constitute more than 50% of annual total supply in the United States. There is no import duty and no quota placed on the amount imported.

The National Marine Fisheries Service has indicated that shrimp consumption is less sensitive to price changes than is beef and pork consumption. They predict that the annual per capita shrimp consumption will exceed 3.0 pounds before 1980. They feel that the fraction of shrimp supply imported will have to increase to meet the added demand. By 1980, world shrimp demand will equal the world's estimated harvest potential, and they feel that beyond 1980, aquaculture will have to supplement world supply in order to continue to meet world demand (Figure 4.7).

In general it is speculated that the dollar value of fishery imports will rise faster than the annual tonnage imported, because a greater fraction will be high-value species. Domestically cultured fish products can be substituted for some of these imports, provided the operation is economically viable.

Some food market analysts predict a growth in U.S. fish consumption through development of a new aquacultural industry based on advanced technology. This has occurred in the chicken broiler industry. For the 30 year period 1939 to 1969, per capita consumption of chicken rose from about 1.5 to 35.0 pounds per year, while per capita fish consumption remained at 10 to 12 pounds per year. In modern broiler technology, food conversion ratios improved from somewhat less than 5 pounds of feed per pound of flesh to about 2 pounds per pound.

FIGURE 4.7: MARKET PROJECTION ON FUTURE U.S. SHRIMP CONSUMPTION (HEADS-OFF SHRIMP)

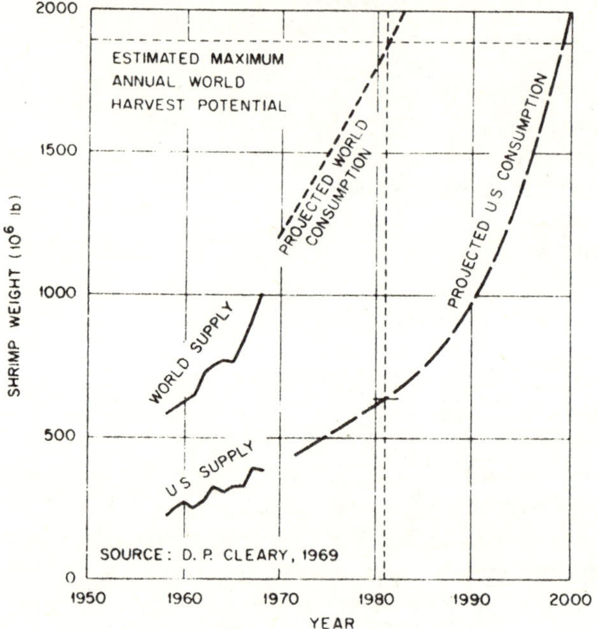

Source: ORNL-4797, July 1972

Cultured species like catfish and rainbow trout under adequately controlled environments do convert nutritionally balanced feed to flesh as efficiently as in broiler production or better. For some other species like shrimp, food conversion efficiency is low because suitable food formulations have not yet been developed, and only natural foods like low-cost fish can be fed at this time.

Food formulas are being evaluated now, and with other improvements including environmental control, a cultured product that is superior to the corresponding wild species could significantly alter the per capita consumption of the fish foods in the future.

Potential for Heat Utilization

Fish culture facilities may be located at power plant sites to utilize land area surrounding the power plant. Power and water are available to blend water streams to achieve water temperature control. Flowing-stream thermal aquaculture may permit year-round intensive culture of some species, an improved product quality over that cultured in a pond on a seasonal basis, and a significant reduction in the costs of culture.

Estimates on the growth of thermal aquaculture in relation to waste heat are difficult. Few demonstration projects are available, and yield data are scarce. One may gain insight into the relationship between heated water availability at power plant sites and the potential for aquaculture from the following assumptions.

Water Utilization: About 2,000 MW of waste heat is generated for each 1,000 MW of electricity produced. About 1,000 million gallons per day (mgd) of cooling water is required to dissipate this waste heat with a 20°F rise in water temperature. If the average ambient temperature of the inlet water is assumed to be 50°F for the colder half of the year and 70°F for the warmer half, and if 70°F is the temperature to be maintained for best growth, then heated effluent (at 70°F) would only be used for thermal aquaculture during the colder half of the year.

During the warmer half, ambient temperature water at 70°F would be used instead of the heated effluent at 90°F. Thus heated water would be used only half of the year. Even during this period the heat is not "consumed", and "thermal pollution" is not reduced significantly.

Fish Yields: The 1,000 mgd (700,000 gpm) of water is distributed over 1,000 acres of working water surface. This is about one-half the size of an exclusion area for a 1000-MW(e) nuclear power plant.

At a fish yield of 10 tons/acre-year (a pessimistic value for intensive culture), an annual production of 4,000 tons or 0.02 pound of fish harvested per 1,000 gallons of H_2O could be realized. This is less ambitious than the production rate at Thousand Springs Trout Company in Buhl, Idaho, where 60,000 gpm of water is distributed over 10 acres, and a harvest of 100 tons/acre-year or 0.06 pound of fish harvested per 1,000 gallons of H_2O is achieved.

For a U.S. population of 200 million and a per capita fish consumption of 10 lb/year, the national consumption of fish food would be 2,000 million pounds. If 10% of this fish consumption were supplied by thermal aquaculture, the equivalent of twenty-five 1000-MW(e) power plant installations of the type postulated would be required. If per capita consumption increased to that in Japan (100 lb/year), the number of 1000-MW(e) power plant aquaculture installations needed would be 250. In terms of land requirements, 10,000 to 100,000 acres would be used.

It is, of course, extremely difficult to predict a market for a new technology like termal aquaculture, and a thorough market analysis is required. Further, the impact of thermal aquaculture on waste heat utilization should be considered on a site-by-site basis, because water quality is highly variable and the ambient seasonal temperature of water used for cooling purposes is important.

Conditions in one section of the country may not apply to other sections. Even within a region, the temperature and quality of waters are highly variable. If, for example, water temperatures are lower in the winter than for the simplified case presented, then fish productivity would be adversely affected. Therefore, generalized projections on a national basis can be very deceptive.

Technological Problems and Development

The utilization of waste heat for aquaculture will have little effect on the amount of thermal energy to be dissipated. However, the waste heat can be used to increase food production. In some instances, fish production would result in a reduction in discharge temperature, since ambient temperature water would be blended with the warm effluent to maintain the optimum temperature range for fish growth. In this case the temperature of the return water stream would be reduced. Thermal plant cycle efficiency would only be affected if the winter discharge cooling water temperature were maintained above normal values.

Only once-through cooling has been studied to date. Aquaculture in conjunction with closed-cycle cooling towers has the advantage of higher available water temperatures but would require a feasibility study, because tower blowdown rates (which would remove wastes) are at least 20 times less than for a once-through system.

The effect of particulates and increased dissolved solids in the blowdown as well as biocides added would have to be considered. Fish culture in the main recirculation stream could be a possibility, but fish wastes will have to be treated prior to recirculation to the power plant condenser. The adaptation would require further study.

Another possibility for using warm water from a cooling tower system is to circulate the water through a heat exchange system (such as that described earlier for greenhouses and animal shelters) to maintain the temperature of a building which houses aquaria or fish culture tanks.

In this enclosed concept, already in the demonstration phase, large tanks are stacked vertically on frames. Each tank contains sufficient water for 500 one-pound fish. The water is recirculated continuously through the tanks to filters and aerators. The amount of heat required to maintain the building temperature depends on the building surface area, insulation, and climate, but would be in the range 0.25 to 0.5 MW per acre of space used.

Large-scale use of waste heat for aquaculture would probably not be considered until demonstration projects at existing sites indicate an economic viability. The projects mentioned earlier may serve this purpose.

Since the demonstration phase may occupy several years, it is unlikely that larger facilities will be planned soon for plants under construction or design. Although such facilities could be installed at a later time, it would be preferable to include the aquaculture facility in the original site selection and planning.

Engineering design and evaluation are needed for intensive aquaculture systems. Applied research and development work would be necessary to complement engineering tests. For a given species, mass culture techniques can be quite different from laboratory experiments. Flow rates for channel culture must be optimized so that energy spent on physical activity is minimized and food energy conversion into flesh is maximized. Aeration systems should be evaluated.

Fish handling devices for transferring and harvesting in a flowing system need to be considered. Fish waste treatment systems need to be designed and potentially represent a significant problem. For the near term, wastes might be diluted by installing relatively small aquaculture farms at each power station, thus holding waste concentrations low, consistent with water quality standards. Low-cost nutritionally balanced feeds must be made available. Selective breeding should be considered to produce species particularly amenable to intensive culture. Fish culturists must be able to furnish fingerlings the year-round in order to have truly continuous culture. Medicinal treatment methods must be available to treat fish diseases rapidly, particularly in intensive culture. Water quality must be satisfactory. Other technical and nontechnical problems may include the following.

(1) To increase the reliability of heated discharge water, it may be necessary to practice aquaculture at multiple-unit plants. Only a fraction of the total volume of heated discharge water would be used for aquaculture, so that in the event of an outage, a switch could be made from a nonoperating to an operating unit.

(2) Even if multiple units are available, unprogrammed shutdowns could cut off the warm water supply suddenly. Such rapid temperature changes could be lethal, and, at least, fish growth rates would be lower until the power plant resumed operation. However, it might be necessary to provide for rapid valving to an alternate operating unit or an auxiliary supply or to stop the water inflow so that thermal shock is minimized as a result of the shutdown. Systems with large thermal inertia would be less affected. Sudden temperature changes on startup could be ameliorated also by gradual blending of heated discharge water with recirculated ambient water.

(3) Batch chlorination of coolant water may result in a residual free chlorine concentration that is toxic; this may be prevented by aeration to drive out the gas, by reverting to continuous chlorination instead of the conventional batch treatment, or by substituting mechanical cleaning devices or periodic thermal shock treatment of the cooling tubes of the condenser.

(4) Increased copper concentrations occur in the discharge water from power plants when condensing temperatures above 100°F are employed. Copper tends to concentrate in oysters and causes a green coloration. Copper may not be a problem if the condenser steam temperature is held below 100°F.

(5) Nuclear plant thermal water used for aquaculture must be protected from radioactivity being discharged into the stream. Monitoring of activity in the cooling water would certainly be required. Fossil-fired stations would of course not have this requirement.

(6) Fish wastes discharged from an intensive culture facility may have

to be removed by acceptable waste treatment methods to minimize the BOD discharged to receiving waters and to meet water quality standards. The waste treatment plant size, design, and economics will have to be studied for each facility.

(7) Legal and regulatory restrictions such as water quality, water rights, and prior appropriation (regulations on the total amount of water usable in a power plant) may influence the viability of the idea in certain regions, including many western states.

(8) Regulatory restrictions on the discharge of heated water may eliminate the once-through approach that has traditionally been used in hatcheries.

(9) Insurance costs might have to be borne by a food cultivator to cover damages that might result from a sudden accidental release of radioactivity or chemicals from a nuclear or fossil power plant, a statistically low possibility but a real one.

Various types of integrated systems may be considered. Multispecies culture systems might be considered, including finfish in channels, conversion of fish wastes to algae, and intensive oyster culture fed on this algae.

Agriculture-aquaculture systems might be considered, particularly in the summetime when thermal effluent temperatures may be too warm for fish culture. Greenhouses might be used as cooling towers to extract heat from thermal effluents, and the discharge from greenhouses may be used for fish culture. This integrated system might permit the maximum utilization of waste heat for food production, and simultaneously incorporate aquaculture into a closed recirculating system instead of a once-through cooling system. However, fish waste treatment would be a necessary part of this system and may be expensive.

Demonstration of intensive culture using a culturable fish species and power plant thermal effluents is needed, and information is needed on the degree to which yields are improved by waste heat utilization in small pilot systems. The facility at the Gallatin Steam Plant should answer some of these questions for that specific site and species; work now being carried out by Long Island Oyster Farms, Inc., at Northport, New York, will provide additional information on oysters, clams, and scallops; and work in Florida, California, and Maine should provide information on other species.

Additional demonstrations at other sites for other species, however, are still needed. Once the data are obtained, sufficient information will be available to determine the incentives for performing the engineering, biology, and chemistry necessary for thermal aquaculture on a commercial scale.

Summary

Thermal aquaculture is a method for using heated effluents productively, but it does not necessarily reduce the heat disposal problem of the power plant. Basic data show that warm-water fish growth rates could be increased by a factor of 2 to 3 by controlling the temperature of water medium within the range 75° to 85°F. Yield potential can be optimized in flowing-stream aquaculture, employing nutritionally balanced feed and oxygenation of the water. Food conversion efficiency also improves with temperature control.

With technical innovations, a significant reduction in production cost comparable to that already achieved in the chicken broiler industry could occur. Seafood consumed in this country is largely wild stock, and comparatively little effort has been expended to culture fish on an intensive basis as is done with land animals.

Waste heat is unlikely to be used in large-scale applications until successful demonstrations have been achieved. Therefore, the short-term impact of this activity on power plant siting should be small. Over the longer term, however, the possibilities for aquaculture should be considered during site selection. Site-oriented demonstration programs are needed to provide the technical data that will indicate the extent of improvement in quality and yield of culturable fish species through greater environmental control. These demonstration programs, some already in progress will show the viability of thermal aquaculture.

Thermal aquaculture will not diminish the amount of waste heat to be rejected from a power plant. During the summer, to maintain optimum growth temperatures, it may be desirable or necessary to dilute the heated discharge water with ambient temperature water. To the extent that ambient temperature water for blending purposes is available, this dilution process will reduce the temperature of the water discharged to the receiving water body.

The cost of this dilution would be borne by the aquaculture operator and the power producer. During the winter, ambient temperatures may not be as warm as desired, and this will reduce growth.

Unless it is removed from the culture stream effluent, fish waste could contribute to the pollution of the receiving water by increasing the biological oxygen demand. The cost of adding a treatment plant to take care of fish wastes, particularly in the effluent of an intensive culture facility, should be considered and evaluated as part of the economics of thermal aquaculture.

Waste Heat Utilization in Wastewater Treatment

ARIZONA UNIVERSITY STUDY

Introduction

Municipal and industrial wastewater effluents are growing rapidly. They are, however, increasingly being looked upon, not as pollutants but as a secondary water resource. The use of treated wastewater effluent for cooling in power plants would be a positive contribution in water conservation.

The productive use of waste heat in wastewater treatment can be mutually beneficial. A combined power plant-wastewater treatment system would generate recyclable water for other uses in addition to being used for the heat rejection process. This research effort was performed by R.A. Sierka and R.A. Fazzolare of Arizona University for the Office of Water Resources Research and was published as PB 226 141, April, 1973. It focused on physical-chemical methods of waste treatment at elevated temperatures.

System Concept

After extensive laboratory investigations of alternatives in treatment process using various chemical additions at elevated temperatures, it was concluded that the one-step settling-adsorption process utilizing ferric chloride ($FeCl_3$) and powdered activated carbon (PAC) at temperatures between 40° and 60°C would effectively remove the COD, suspended solids, turbidity, and sulfates to levels of conventional

FIGURE 5.1: POWER PLANT–WASTEWATER TREATMENT COOLING LAKE COMPLEX

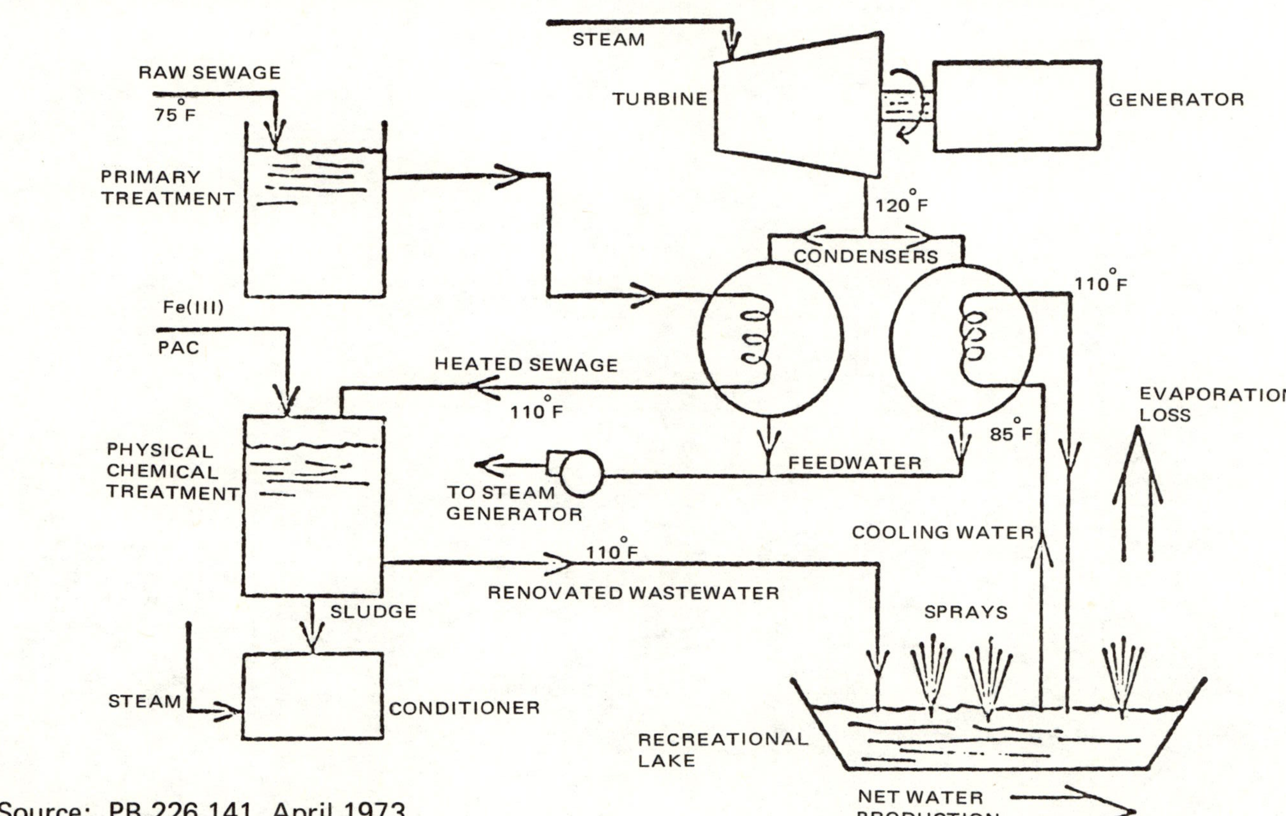

Source: PB 226 141, April 1973

FIGURE 5.2: POWER PLANT–WASTEWATER TREATMENT COOLING TOWER COMPLEX

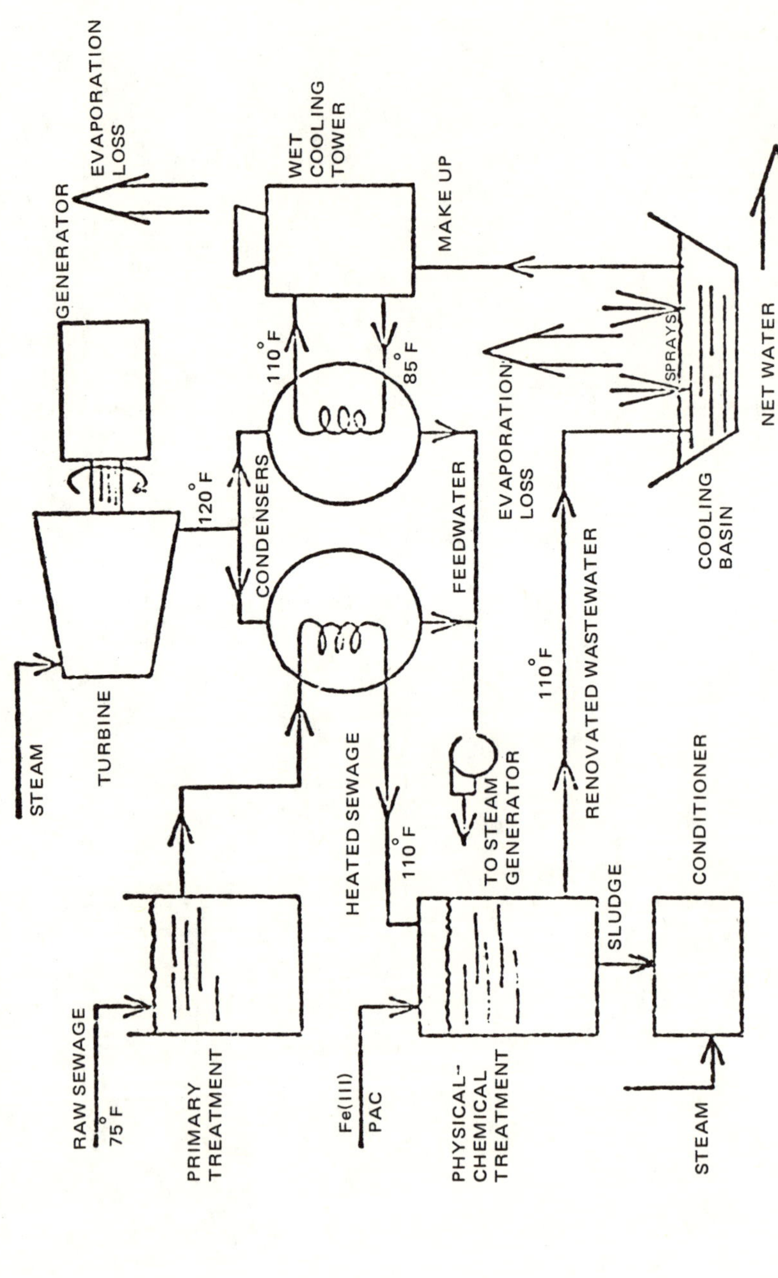

Source: PB 226 141, April 1973

secondary treatment. Two possible system flow diagrams for a combined power plant-wastewater treatment complex are shown in Figures 5.1 and 5.2. The raw sewage is passed through a primary settling facility to remove large particulates that might affect heat transfer operation. The primary treated wastewater is then passed through a special condenser to accommodate the characteristics of the fluid where it is heated to approximately 110°F (43°C).

The back pressure of the turbine will be about 3.5" Hg to permit 120°F steam in the condenser resulting in a minor loss of thermal efficiency. The chemicals are added to the heated wastewater where adsorption, coagulation, flocculation, and settling occur. The treated effluent can then flow into a lake as shown in Figure 5.1.

The lake serves as a cooling pond which may be equipped with spray systems to accelerate the cooling process and minimize surface area. Cooling water is withdrawn from the lake to provide for the additional heat rejection in a conventional condenser.

Alternatively, in Figure 5.2, the renovated but warm wastewater is sent to a outflow channel, which can act as a cooling area, if equipped with a spray system, prior to its ultimate reuse. The additional cooling is provided by a wet cooling tower with renovated wastewater being used as make up. Sludge from the physical-chemical treatment is removed and dewatered with the aide of process steam available at the plant.

Experimental

The viability of the above systems were tested experimentally by using settled primary sewage obtained from the wastewater collected at the City of Tucson Wastewater Treatment Plant which serves a population of approximately 300,000 and is almost exclusively a domestic waste.

Table 1 lists the average chemical and physical properties of the plant influent and primary settled effluent. Experiments were conducted in the 24° to 75°C range by placing six, 500 ml beakers in a water bath which maintained temperatures within 2°C of the desired value.

Mixing was accomplished by a six-position Phipps and Bird stirring apparatus. Ferric-chloride ($FeCl_3$) in the 0 to 150 mg/l range as Fe(III) and Filtrasorb-400, a powdered activated carbon (PAC) in concentrations up to 300 mg/l served as metal coagulant and adsorbent respectively.

TABLE 1: CITY OF TUCSON—AVERAGE INFLUENT AND PRIMARY EFFLUENT CHEMICAL AND PHYSICAL ANALYSES

	Wastewater Influent 1971–72*	Primary Effluent May, 1971*
Total solids, evaporation (105°C)	941	--
Suspended solids	219	71
Dissolved solids	722	--
Fixed residue	601	--
Volatile residue	340	69
Total alkalinity ($CaCO_3$)	282	--
Total acidity ($CaCO_3$)	22	--
pH	7.5	7.4
BOD	211	--
COD	282	237
MBAS	6.4	--
Grease	79	--
NH_3-N	24.6	23.7
NO_2-N	--	--
NO_3-N	--	--
Organic-N	17.8	8.7
Total-N	42.4	32.4
Silica (SiO_2)	50	--
Aluminum & Iron (Al_2O_3 & Fe_2O_3)	33	--
Iron (Ferrous)	1.1	--
Total Iron	1.5	--
Calcium	93	--
Magnesium	24	--
Hardness	427.5	--
Chloride	103	--
Carbonate	0	--
Bicarbonate	344	--
Phosphate, total	--	41
Phosphate, ortho	34	--
Sulfate	166	--
Sodium & Potassium	97	--

*All values measured as mg/l, except pH

Source: PB 226 141, April 1973

A number of investigations regarding the sequence of addition of PAC and $FeCl_3$ to the wastewater were conducted including: (1) mixing PAC and $FeCl_3$ together and permitting it to stand for five minutes before addition; (2) adding $FeCl_3$, rapid mixing for two minutes then adding PAC; (3) adding PAC, mixing at 30 rpm, which is the minimum rpm necessary to keep the carbon in suspension, for five minutes followed by $FeCl_3$ addition and rapid mixing at 100 rpm for two minutes and (4) simultaneous addition of PAC and $FeCl_3$ without prior mixing followed by mixing at 100 rpm for two minutes.

Methods (1), (2) and (3) provided essentially identical results, while method (4) was relatively inefficient. From an operation standpoint method (3) was chosen and retained throughout the investigation as were the conditions of 30 rpm flocculation for 40 minutes followed by 20 minutes of quiescent settling. In some runs samples were taken as a function of time and the appropriate performance parameter measured. *Standard Methods for the Examination of Water and Wastewaters,* 13th edition, was followed for all analysis.

Experimental Results – Chosen as performance parameters to study the effect of heat in the physical-chemical treatment system employing $FeCl_3$ and PAC was COD, turbidity and phosphorus.

Turbidity removal as a function of temperature was rated with varying doses of $FeCl_3$ from 10 mg/l through 150 mg/l as Fe(III). Turbidity was measured in the Hach Turbidimeter–Model 2100-A as FTU. The results indicate a definite change in removal pattern with Fe(III) dose.

Heat generally beneficially affects turbidity removal below 60 mg/l Fe(III) doses, however, at 60 mg/l the reverse is true; it has a negligible effect at 90 mg/l and beneficial effect at 120 mg/l and 150 mg/l of Fe(III).

The addition of 300 mg/l PAC to the $FeCl_3$-wastewater system markedly improves the turbidity removal. However, again at the 60 mg/l Fe(III) concentration there is a deterioration of turbidity removal with increased temperature. Improved system performance is explained by the increased number of nucleation sites which the carbon particles present to increase the rate of flocculation.

Wastewater pH was plotted as a function of iron dose at temperatures from 24° to 74°C. At Fe(III) concentrations in excess of approximately 90 mg/l treated water will require pH adjustment.

Visual observation throughout these experiments showed floc characteristics and settling rates enhanced by increased temperature. Accordingly, a series of runs were made to delineate the magnitude of the temperature effect. Data was presented for 90 mg/l Fe(III), in wastewater with turbidities measured with settling time.

These data pointed out the beneficial effect of temperature in decreasing turbidity to less than 2.0 FTU within six minutes of settling time at reaction temperatures between 45° and 50°C. When 300 mg/l of PAC is added to the 90 mg/l of $FeCl_3$ a still further enhancement in settling times and reduced turbidity level was seen.

Sludge volume relationships with temperatures for the $FeCl_3$-PAC system were investigated. Sludge volume decreases with temperature increases are noted for each level of PAC (0 to 300 mg/l) at three $FeCl_3$ doses; 60, 90, and 120 mg/l. The relative sludge compaction, as determined by laboratory tests with 500 ml volumetric cylinders, is not representative of actual compaction that would be found in secondary clarifiers. Nevertheless, the experiments demonstrated that as temperature increases from 24° to 74°C sludge volume decreased by an average of 25%.

The granular character of the sludge gave a very stable floc. Even upon agitation of the settled sludge any reentrained solids were quickly resettled. This behavior led to the experiments concerned with rates of filterability.

Two runs were performed at 24° and 60°C for a wastewater treated with 90 mg/l of $FeCl_3$ and 300 mg/l of PAC. Approximately 25% more filtrate was removed in two minutes at 60°C than 24°C, which would be expected from not only viscosity considerations, but the physical characteristics of the sludge.

Concern with eutrophication of our nation's waters has led to the enactment of many state laws regarding the removal of nutrients from wastewater plant effluents. In these studies, only phosphorus removals were measured. It was seen that operating temperatures above 50°C lead to the very rapid deterioration of phosphorus removal. Phosphorus can be removed, however, rapidly and readily, up to 40°C to levels exceeding 98%.

Heating sewage in the temperature range 24° to 90°C increased the soluble COD (SCOD). Above 60°C rather large SCOD increases were noted. The effect of this increase in SCOD has impact on the physical-chemical system in two ways. First, the adsorption capacity of the carbon is utilized more quickly and second the molecular architecture of the adsorbates has changed. With the application of heat, high molecular weight compounds degrade to lower molecular weight constituents which normally display increased solubility and, therefore, are more difficult to adsorb.

This leads to higher effluent COD residuals. The total COD removed from Tucson primary effluent was studied as a function of temperature and PAC concentration with a constant iron dose of 90 mg/l as Fe(III). It appeared from these experiments that 60°C would be the highest desired operating temperature. The maximum COD removal with 300 mg/l of PAC was found to be 87%. This translates to a residual COD of 30 mg/l which is within the generally achievable goals of secondary biological treatment.

Conclusion

The above study has shown the engineering viability of the concept of a combined power plant–physical-chemical wastewater treatment complex. The physical-chemical mode appears to be attractive in this design configuration since it is less sensitive to hydraulic, organic and thermal variation that invariably must be expected in the operation of a power plant. Power plant downtime can be accommodated by variations in chemical addition.

The proposed combined system will not only relieve the demand on primary water resources for power generation, but also will produce a net high quality water resource.

URS RESEARCH COMPANY

Another investigation to evaluate the technical and economic feasibility of utilizing waste heat to improve wastewater treatment process efficiency was performed by F.J. Agardy, M. Fields, and M. Staackmann of the URS Research Company and was published by the Environmental Protection Agency as PB 217 880, January 1973. The information in the following section is extracted from that report in which particular attention has been given to the grouped utility concept.

Introduction

The treatment of domestic and industrial wastes generally takes place at ambient temperatures due to the large quantities of wastewater involved. However, it is recognized that most physical, chemical and biological processes operate with greater efficiency at elevated temperature (10° to 30°C). Conversely, most processes, particularly those of a biological nature, are severely curtailed at low temperatures, precluding their use in very cold climates (e.g., where wastewater temperatures are in the 0° to 5°C range). Two general alternative solutions are available:

(1) Raising the temperature of the water to a level where adequate biological activity can take place.

(2) Adopting physical and/or chemical treatment methods which are less sensitive to temperature.

The high cost of heating wastewater has traditionally precluded the former. Recognizing that warm water represents a sizable resource of low-cost heat, the possibility of locating utilities (power and wastewater treatment) to maximize heat transfer offers an attractive solution to problems of wastewater treatment in new facilities especially those located in cold climates. In this regard, the group utility concept was considered as a viable solution to the problem.

The Grouped Utility Concept

The grouped utility concept has been presented for other applications in the past and can be defined as the locating and designing of two or more utilities so that they can take advantage of each other's contributions and requirements.

In particular, the grouped utility concept has been considered for use with power plants in which the combination of electric power generation and seawater desalting at the same location is available. For this study, the combination of interest is electric power generation and wastewater treatment.

Utility designs have generally not recognized any interdependence of electric power plants and wastewater treatment plants. Power demands for the future will have to be met by new power plants. The opportunity exists to design future plants to obtain maximum benefits from power generating systems. One possible design would be a combined power/wastewater treatment plant wherein some of the heat rejected from the power generation is utilized to improve efficiency in wastewater treatment. This combined system has the advantage of extracting some benefit from the waste heat before it is discharged to the environment.

Two general types of relationships can be identified when the utilities are combined. The simplest is where one-directional dependency is exhibited. For example, the power plant benefits the wastewater treatment plant but normally operates independently. The second type of relationship is mutual synergism, wherein each utility derives significant benefit from the other.

This could be true if the power plant experienced lower costs due to the elimination of the necessity for cooling ponds or cooling towers. It must be understood that such waste heat utilization does not eliminate the heat, but merely transfers it from one place to another. The total heat load on the receiving water may still be the same, except for whatever heat would be lost to the atmosphere while the heated wastewater passed through the treatment plant.

Besides the benefits associated with waste heat, a grouped utility system might have other valuable outputs such as low cost, off-peak electrical power, off-peak mechanical power, off-peak steam, spent radioactive fuel elements, or perhaps low level radioactive wastes.

In considering sequential use of resource energy and materials, it may be advantageous to consider a grouped utility system with more than two members to maximize total benefits from the combination. The inclusion of other appropriate industrial operations into the combine would depend on the local conditions and resources available.

Heat Transfer Mechanisms

It is important to identify methods of transferring heat into typical wastewater treatment processes. Considerations of heat transfer mechanisms must include the basic methods, the actual equipment to be used, and the potential transfer efficiencies which can be achieved. Each of these areas will be briefly discussed in the following paragraphs.

Whereas the majority of waste heat available is in the form of heated water discharges, hot exhaust gas is also significant for certain industries and locations. However, no discussion on hot exhaust gases will be presented, although heat transfer from gases was studied and mentioned in the original report.

Heat Transfer Methods – The overall problem of heat transfer can be broken down into two phases. The first phase is transfer of heat from the generation source to the point of use, and the second phase is transfer of heat into the appropriate process. Wherever a grouped utility system is established, the first phase of transfer would be essentially trivial or even nonexistent. For current situations, however, this phase becomes very important and must be carefully examined to define areas of limitation.

The various heat transfer methods of interest are direct transport, direct contact, and indirect contact. In direct transport the heat is transferred by means of the hot fluid physically moving from one place to another. This covers, for example, the hot cooling water flowing in a pipeline from the power plant to the wastewater treatment plant. Such transfers normally involve negligible losses of heat, but under conditions of very long pipelines and/or very cold ambient air temperatures, significant heat losses may occur.

In direct contact the heat is transferred from one fluid to another by direct physical contact between fluids. This covers, for example, hot water mixing directly with wastewater to be treated, or hot exhaust gases bubbling up through such wastewater. In indirect contact the heat is transferred from one fluid to another with the fluids being separated by an appropriate physical barrier. An example of this would be the warming of wastewater by hot cooling water in a shell and tube heat exchanger.

Heat Transfer Equipment – For direct transport heat transfer, liquids and gases can conveniently be transported by means of pipelines appropriately sized. An alternative for liquid transfer would be the use of an open channel for flow, but it is recognized that this method greatly increases the associated heat losses to the atmosphere. Pipelines are usually buried underground and can be insulated to further reduce heat losses.

Direct contact transfer equipment can also consist simply of pipelines to direct the fluids to the point of use. In utilizing hot gases the piping should inject the gases well below the liquid surface to maximize contact time. Properly designed diffusers must be used to achieve appropriate flow rates and gas bubble sizes. Two approaches can be considered for use of heated water. Either the hot water may be permitted to flow directly into wastewater to be treated (in which case a dilution occurs at the point of treatment) or the hot water could (after pretreatment) form all or part of the incoming water supply for the community being served. (This latter method avoids dilution, but could result in considerably greater heat losses since this water flows throughout the entire water system).

Indirect contact transfer equipment could include the full range of heat exchangers currently used such as rotating disc, double-pipe, plate and frame, shell and tube, and plate heat exchangers. The two basic types consist of the confinement of one or both fluids. Examples of where only one fluid is confined are:

(1) Heater coils immersed in a wastewater treatment tank with heated cooling water or hot exhaust gases circulating through the coils and the wastewater circulating by natural or forced convection;

(2) The use of a cooling tower or trickling filter type device in which wastewater trickles down through a packing media and hot exhaust gases flow up the packing. (This could also be considered a variation of the direct contact type heat transfer.)

Examples of the system where both fluids are confined include all the arrangements of pipes, tubes, plates, and vessels in heat exchanger types indicated above.

Another unusual heating method should be mentioned although it does not strictly originate by using waste heat. A trickling filter operating in a sewage treatment plant in Nora Springs, Iowa uses a translucent roof which allows radiant energy from the sun to warm the filter on cold days. The concept of using a cover over trickling filters or other wastewater treatment processes may have value if the space over the process can be beneficially heated by use of hot exhaust gases.

Heat Losses and Transfer Efficiencies – Heat losses associated with direct transport of heated water in pipelines were considered in detail for a wide range of conditions. Underground pipelines were examined for a variety of temperatures to determine heat losses by conduction to the surrounding soil and air.

Heat transfer efficiencies for heat exchangers are best expressed by the overall heat transfer coefficients applicable for the various types of exchangers. It was found that in comparing heat exchangers with equal heat transfer areas, units with the largest coefficients are most efficient. The most efficient units tend to be the shell and tube and the plate heat exchangers. It appears that for the specific problems of waste heat utilization, the plate heat exchanger can provide the greatest coefficients and the highest efficiencies of heat transfer.

For plate heat exchangers, it is possible to use values of 1° to 2°C for the exchanger temperature approach (the water to be heated leaves the heat exchanger at a temperature 1° to 2°C below the inlet temperature of the warmer water). If fouling or scaling problems with these heat exchangers become difficult or expensive, other heat transfer methods such as rotating disc and cooling tower or trickling filter heat exchangers may be appropriate.

Potential Problems with Coating/Scaling of Heat Transfer Surfaces – The efficiency of heat exchangers for indirect contact heat transfer depends partially on the maintenance of clean heat transfer surfaces within the equipment. Wastewater will be passing through one side of the heat exchanger and if any coating or scaling occurs in the passages, heat transfer efficiency will be adversely affected.

Careful consideration must be given to the heat exchange method selected and the type of wastewater which is to be heated so that harmful levels of suspended solids and/or dissolved materials can be avoided. It is possible that pretreatment of the wastewater will be necessary no matter what type of heat transfer mechanism is utilized. Shell and tube and plate heat exchangers would be more subject to plugging, as well as more difficult to clean, than rotating disc and cooling tower or trickling filter heat exchangers.

Problems with scaling or fouling can occur with inorganic salts or organic materials. The particular problems will depend on the nature of the wastewater involved. Raising the wastewater temperature can cause a lower solubility for certain salts which could result in scale formation. Such scale can be avoided by use of chemical additives or the scale can be removed by acid cleaning.

Both of these methods are very inexpensive. However, such scale formation is not likely for the small temperature increases involved. Fouling with organic material is, however, quite likely to occur unless proper precautions are taken. Very little information is available concerning fouling and costs to prevent fouling. It should be pointed out that if such costs become significant, they could have an impact on the overall cost/benefit analysis of this study.

Temperature Dependency of Unit Processes and Establishment of Realistic Temperature Ranges

In order to establish definitive temperature ranges and to fill in the real world data gaps for the operation of activated sludge and ancillary systems, actual treatment plant performance data must be examined. In this study informational data sheet was sent to U.S. cities with a population of approximately 100,000 or more and average January air temperatures of 30°F or less. Of the 65 cities contacted, 19 (or 29%) responded with data on their processes. Of these, 10 (56%) contained data on activated sludge, and 9 (47%) showed operational data on other biological systems.

The operational data consisted partially of air and unit process temperature, flow rates, loading rates and effluent quality in terms of the 5-day BOD (Biochemical Oxygen Demand). From these data, operation of processes such as sedimentation and activated sludge could be evaluated to discern what effect various temperature ranges have on the effectiveness of treatment of these and other ancillary processes.

It is known that temperature differences will affect the rate of settling of particles in liquids by causing changes in the viscosity of the liquid and changes in density of both the particle and liquid. Data also indicate that as temperature is increased, the velocity at which particles will settle is also increased. In observing wastewater treatment plants, it was found that greater efficiencies were obtained from the sedimentation tanks in summer than in winter. The cause of this increased efficiency in these cases has been attributed to the higher temperature of the sewage in the summer.

Plots of data were made of temperature vs percent removal of suspended solids. A comparison of temperature data as found in the literature, derived from bench scale models and actual data received from treatment plants was made to determine what correlation (if any) for removal efficiencies can be made with the physical/chemical changes such as density, viscosity, solubility and reaction rates.

It was found that detention time is a factor in the overall removal efficiency of the primary sedimentation process. It was apparent that even at lower detention times (15 to 60 min), high quality effluent could be achieved if wastewater temperatures were raised significantly. The implication is that greater removal of suspended solids is possible and this beneficial effect can be utilized in two ways: (1) existing units could be operated at shorter detention times and still retain good removal efficiencies; (2) new units could be designed with smaller capacities to achieve the same performance. It was also found that an optimum temperature range for primary sedimentation lies between 10° and 20°C for average detention time employed by wastewater treatment plants.

Other data showed there was a greater removal of suspended solids with an increase of sewage temperature. The removal of suspended solids for sewage at average normal temperatures (10°C or 50°F) is between 35 and 65%. There was an incremental increase of 20% removal of suspended solids per each 5°C rise in temperature. This increase is only relevant as long as the normal range of detention periods (45 to 120 min) is maintained.

The most common measure of efficiency in treatment plants is related to the amount of BOD removal. For a primary sedimentation basin operating at average sewage temperatures (10°C), a removal range of 20 to 40% BOD is possible. The field data showed a 35% BOD removal also at the average sewage temperature. This value lies within the reported 20 to 40% range.

Experiments indicated that if temperatures of the sedimentation basin could be maintained at 2.5°C above normal sewage temperature, the upper limit of normal removal efficiency of BOD (40%) would be attained. With an increase of 10°C, the removal of BOD could be increased by 15% above the upper limits of the BOD removal range. Since the overall efficiency of biological processes is linked with pre- and final sedimentation, the increase of the presedimentation unit would have an impact on the total treatment system.

Temperature Effects on Oxygen Utilization – It is apparent that temperature influences the rate of chemical and biochemical reactions. It can be shown that in the ranges of optimum biological activity, a two- to three-fold increase in reaction velocity is experienced for each 10°C rise in temperature. Since oxygen utilization represents a sequence of related reactions (oxygen transfer, adsorption, oxidation, etc.), it is probable that temperature effects are not the controlling factor under all temperature conditions.

In most activated sludge treatment plants air is supplied in excess so as to ensure an adequate supply of oxygen required for biological metabolism and a high level of mixing. It is doubtful, therefore, that an increase of temperature would cause any change in oxygen utilization with this existing operation of supplying excess air.

Effects of Temperature on Nitrification – One of the prime nutrients which contributes to the eutrophication of surface waters is nitrogen in the fixed forms of ammonium and nitrate ions. Many states are now requiring that Nitrogen Oxygen Demand (NOD) be considered as part of the analysis performed to ascertain the total pollutional impact on these surface waters.

Nitrogen may be present in a wastewater as organic nitrogen, ammonia nitrogen, and nitrite and nitrate nitrogen. During the course of biological treatment, organic nitrogen is broken down to ammonia and, depending on the process conditions, ammonia may be progressively oxidized to nitrite and nitrate. Some organic nitrogen forms are refractory and pass through the process.

During the BOD removal phase of the process nitrogen is assimilated into bacterial protoplasm. A portion of this nitrogen is released back to solution through the endogenous metabolism of the biomass. Nitrites and nitrates are biologically reduced to nitrogen gas under appropriate conditions. It is usual that conventional activated sludge and trickling filter processes remove 20 to 40% of the total nitrogen.

The effects of temperature on nitrification in the activated sludge process is shown in Table 2, which indicates the effect of increased temperatures on some of the nitrification parameters.

TABLE 2: TEMPERATURE EFFECTS ON NITRIFICATION PARAMETERS

Temperature, °C	Sludge Age (days)	Required Aeration Volume ($ft^3 \times 10^4$)	Hydraulic Detention Time (hr)	Oxygen Requirement in Addition to Carbonaceous Oxidation (lb/day)
10	10.5	7.67	4.6	1740
15	5.3	5.82	3.5	1530
20	3.5	4.07	2.44	1388

Temperature Effects on Disinfection – Disinfection is a term applied to those processes in which pathogenic microorganisms are destroyed. Chlorine and chlorine compounds are the main disinfecting agents used in the United States for the treatment of municipal water supplies and wastewater due to the specific high toxicity of chlorine for microorganisms responsible for water-borne diseases.

The primary factors which govern the bacterial efficiency of both free and combined available chlorine are based on the temperature of the water in which the contact is made (the higher the temperature, the more effective the disinfection), and the time of contact of organisms and bacterial agent (the longer time, the more effective the disinfection).

Experiments have shown there are two possible benefits that can be realized from operation at optimum temperature ranges (15° to 20°C). For a given detention time and pH the amounts of residual chlorine needed are reduced by 20 to 40%, hence the applied dosage for combined and free chlorine is also reduced 20 to 40%.

In addition, if the dosage is held constant, then the contact time required to obtain the needed kill is also reduced 20 to 40% with increased temperature. This would allow for construction of smaller contact chambers with a reduced capital expenditure for the chlorination chamber.

Temperature Effects on Activated Sludge

The effects of temperature on the activated sludge process have received little attention in the past because of the lack of an economical method for changing or controlling wastewater temperatures. With larger quantities of waste heat being discharged from various power generation sources and industrial complexes and improved methods for heat exchange and transfer, it may now be possible to improve the efficiency, and therefore, the economic operation of secondary treatment systems.

This improvement due to the utilization of waste heat would be most beneficial in areas which are being newly developed (for optimum location of treatment factilities to waste heat discharge) and in those areas with prolonged cold seasons.

There are many variables which affect the performance and efficiency of the activated sludge process. Some of these are organic loading, temperature, type of substrate, and organism growth pattern. Activated sludge, being a biological process, should have a definite temperature range at which organism growth patterns become optimum. Changes in the growth pattern could affect the competitive position of a species and their relative distribution in a population. This change in population (either increasing or decreasing) due to the temperature would affect the oxygen utilization and the synthesization of organic matter with a definite effect on performance.

The growth of cells and removal of BOD (as affected by temperature) approximately doubles for each 10°C. As the temperature is increased more substrate organic material is required to synthesize 1 gram of solids; in other words, increased utilization of organic matter (BOD) takes place.

The temperature effects on the activated sludge process can be summarized as follows: BOD removal and oxidation is affected by changes in growth patterns. The amount of oxidation depends on diffusion of oxygen into the biological floc. At conventional mixing intensities, relatively large flocs are generated and at low temperatures a low oxygen utilization rate permits diffusion of oxygen to a greater depth in the floc and, therefore, a large portion of the floc is aerobic.

At high temperatures, the increased respiration rate causes more rapid depletion of the oxygen and only a small portion of the floc is aerobic. At high mixing intensities, such as found in an aeration basin, the smaller floc sizes may be fully aerobic under all temperature conditions. At the higher food-microorganism ratios, the dispersed filamentous growths will exhibit a high temperature dependency.

Heat Transfer Costs

The costs of achieving heat transfer from a waste heat source to a wastewater treatment process can be broken down into two basic elements. The first is the cost to transport the waste heat from the source to the location of the treatment plant. This cost is dependent on the distance between the source and the use point and the amount of waste heat transported. The second cost element is that required to transfer heat from the treatment plant into the actual process.

Costs for waste treatment plants are usually expressed in cents/1,000 gallons treated. Costs for heat transfer will also be expressed in cents/1,000 gal of wastewater treated. Heat transfer costs are those costs required to raise the wastewater temperature by a fixed amount. The standard temperature increase of greatest interest for this study is about 10°C, which corresponds to the average temperature increase expected from using electric power generation cooling water and the most efficient heat transfer methods.

Heat Transport Costs – Costs have been calculated for transporting sufficient heated water to raise the wastewater temperature by 10°C in treatment plants with capacities from 1 to 100 MGD. Distances of from 1 to 25 miles between the power plant and the treatment plant have been considered.

The cost items involved include: capital costs for the pipeline and pumping system, operating and maintenance labor for this system, and electric power requirements for pumping. Capital costs have been converted to annual costs based on 4½% for 25 years (typical value used in previous EPA studies).

For any particular pipeline length and flow capacity, costs for this system depend on the discharge pressure of the cooling water, the pressure at which this water must be delivered, and the difference in elevation between the power plant and the treatment plant. It is assumed that the water is available and delivered at ambient pressure (0 psig). Minimum costs occur where the treatment plant is downhill from the power plant and no pumping is required. The worst case considers that the treatment plant is 100 ft above the power plant in elevation.

Using the above assumptions, costs adjusted to 1971 for the heat transport portion have been calculated and are presented below.

TABLE 3: HEAT TRANSPORT COSTS (In Cents/1,000 Gal*)

Distance from Power Plant to Treatment Plant (miles)	**Capacity of Treatment Plant (MGD)**		
	1	**10**	**100**
1	3.7 - 7.3	1.5 - 3.9	1.3 - 2.8
5	8.9 - 13.6	2.9 - 5.4	2.3 - 3.8
25	35.0 - 44.9	9.5 - 12.7	7.0 - 9.1

*The range of cost values covers the cases from minimum cost (no pumping required) to maximum for this study (100 ft of elevation difference between plants).

The costs in Table 3 do not reflect any credit which may be due to the system because of the elimination of the normal piping system used to transport the discharged cooling water to the nearest receiving water. This savings could be significant for certain power plants. Heat transport costs will be negligible in the case of an integrated complex where the power plant is located adjacent to the wastewater treatment plant.

Heat Transfer Costs – The costs associated with transferring heat into any treatment process depend on the amount of heat transferred and the method of heat transfer. The two basic heat transfer methods of interest are:

(1) The direct method, whereby the hot water is added directly to the wastewater. For this method it must be remembered that contaminants in the hot water could be detrimental to certain wastewater treatment processes.
(2) The indirect method, whereby the wastewater is heated by the hot water in a heat exchanger.

The direct method has the advantage of optimizing the total heat transfer, but a dilution of the wastewater occurs and larger volumes of water must be treated. The indirect method has the advantage of raising the wastewater temperature the most, but costs are higher.

The standard case assumed for this study is the one in which wastewater temperature is raised to 10°C, and there is a 1°C temperature approach for the heat exchanger. This requires use of the indirect heat transfer method with the highest efficiency heat exchangers. If the wastewater temperature is originally 10°C, then the standard case means raising it to 20°C.

For this case, the original hot water temperature is 21°C as it enters the heat exchanger. Use of a plate heat exchanger would then yield an exit temperature of 16.8°C for the hot water stream. A heat balance for this standard case reveals that 2.4 gallons of hot water are required for every gallon of wastewater heated. This standard case gives the highest heat transfer costs. Lower costs can be achieved in two ways. Less efficient heat exchangers can be used which give reduced costs but lower temperatures for the wastewater. The second method of reducing costs is to use direct heat transfer. This gives greatly reduced costs, but lower temperatures result and larger volumes of wastewater must be treated.

The cost items involved include capital costs for the pumping, piping, and heat exchanger system, operating and maintenance labor for this system, and electric power requirements for pumping. Costs (adjusted to 1971) for the heat transfer portion using alternative methods have been calculated and are presented below.

TABLE 4: HEAT TRANSFER COSTS* (In Cents/1,000 Gal)

Heat Transfer Method	Capacity of Treatment Plant (MGD) 1	10	100
Indirect: 10°C rise	2.4	1.2	1.1
Indirect: 8°C rise	2.0	1.0	0.9
Direct: 8°C rise**	0.63	0.34	0.28

*Without transport costs (if any).
**The wastewater volume is 3.4 times as great for this case as for the volume to be treated using indirect heating.

Overall Heat Transfer Costs — To obtain the overall costs, the transport and transfer costs are added together. It has been assumed that transport costs for the integrated complex are zero. Overall heat transfer costs for the parameters studied are presented below.

Distance from Power Plant to Treatment Plant (miles)	Method of Heat Transfer	Temp Rises (°C)	Capacity of Treatment Plant (MGD) 1	10	100
0	Indirect transfer	10	2.4	1.2	1.1
	Indirect transfer	8	2.0	1.0	0.9
	Direct transfer	8	0.63	0.34	0.28
1	Indirect transfer	10	6.1-9.7	2.7-5.1	2.4-3.9
	Indirect transfer	8	5.7-9.3	2.5-4.9	2.2-3.7
	Direct transfer	8	4.3-7.9	1.8-4.2	1.6-3.1
5	Indirect transfer	10	11.3-16.0	4.1-6.6	3.4-4.9
	Indirect transfer	8	10.9-15.6	3.9-6.4	3.2-4.7
	Direct transfer	8	9.5-14.2	3.2-5.7	2.6-4.1
25	Indirect transfer	10	37.4-47.3	10.7-13.9	8.1-10.2
	Indirect transfer	8	37.0-46.9	10.5-13.7	7.9-10.0
	Direct transfer	8	35.6-45.5	9.8-13.0	7.3-9.4

Heat Transfer Benefits

The benefits accruing to the various wastewater treatment processes should ultimately be expressed in terms of savings in cents/1,000 gallons treated. In these terms, the benefits may be compared with costs and the benefit/cost relationship for waste heat application may be evaluated. In general, the range of plant sizes from 1 to 100 MGD are explored. It should be indicated that the quantified benefits are based solely on the temperature effects and not the increase in plant size. There are two basic situations to be considered: the waste heat as applied to an existing plant and the waste heat as applied in a new plant.

In a new plant, use of the waste heat can be incorporated directly into design considerations and the most efficient utilization of the heat will result. Incorporation of waste heat application into an existing plant requires the solution of special problems depending on whether direct or indirect heat transfer is employed. If the direct heat transfer method is used, there must be sufficient hydraulic capacity in the system to handle the additional water.

In addition, a suitable point to add the hot water must be located where the combined waters will mix adequately to give a uniform temperature. If the indirect heat transfer method is used, a suitable point to divert wastewater from the main flow path must be located and sufficient area must exist for the heat exchanger equipment. In addition, the hot water pipeline must be situated within the existing equipment network to feed the heat exchanger and discharge to an appropriate receiving water.

Benefits for the new plant will generally appear as lower capital costs since it is possible for smaller treatment units to yield the desired effluent quality. Benefits for the existing plant cannot be so directly quantified since the usual result will be a higher quality effluent than that of a plant not utilizing waste heat. This latter case can be handled in two ways:

(1) It can be assumed that the higher quality effluent will be mandatory in the future and the existing plant (without use of waste heat) would have to add processing equipment (with resultant costs) to meet this goal.

(2) Instead of yielding a higher quality effluent, the use of waste heat could permit a higher flow rate through a process unit at the normal level of quality (in this way the waste heat use saves on added equipment for the case where higher flow rates become mandatory in the future).

Each of the treatment processes will be examined separately to determine the benefits on an individual basis. It appears that the optimum use of waste heat is to raise wastewater temperatures at the entrance to the treatment plant and yield the summation of benefits from each of the processes. Owing to the problems and costs involved with heat exchanger fouling, it may, however, be more advantageous to heat the wastewater at a point further downstream in the treatment plant (e.g., after primary treatment).

Primary Sedimentation — The effect of temperature on primary sedimentation has been discussed previously. By means of a literature examination, the range of detention times from 45 to 120 minutes was investigated to find the effects of waste heat use in the following forms: (1) the reduced capacity permitted for new plants; and (2) the additional capacity required in existing plants if waste heat were not employed (this additional cost is saved in the case of waste heat use). The results of this investigation are given below.

TABLE 5: EFFECTS OF WASTE HEAT UTILIZATION ON PRIMARY SEDIMENTATION

		New Plants	Existing Plants
Capacity required without waste heat use		1.00	1.19 - 1.33
Capacity required with use of waste heat			
Indirect heat transfer:	10°C rise	0.70 - 0.77	1.0
Direct heat transfer:	7°C rise	2.7 - 3.0	3.2 - 3.6

It can be seen that the indirect heat transfer method is advantageous for new and existing plants. The direct heat transfer method is not beneficial in either case since the added volume in the system is not compensated for by reduced capacity requirements at elevated temperature. It must be remembered that the direct heat transfer case assumed was for hot water at a temperature 10°C above the wastewater temperature.

A more detailed study revealed that direct heat transfer only became beneficial over normal operation at hot water temperatures greater than 30°C above that of the wastewater.

The benefits of the indirect heat transfer method indicated above can be converted to cost figures by considering capital, operating, and maintenance costs for primary sedimentation. The resultant savings to this process are given in the table shown on the following page.

TABLE 6: COST SAVINGS FOR PRIMARY SEDIMENTATION
(Heat Transfer Costs Excluded)

Cost Savings in Cents/1,000 Gal	Treatment Plant Capacity (MGD) 1	10	100
New plants	0.18 - 0.23	0.19 - 0.25	0.19 - 0.24
Existing plants	0.15 - 0.25	0.16 - 0.28	0.16 - 0.26

Secondary Clarification — The effects of temperature on secondary clarification are similar to those for primary sedimentation. Since the same mechanism is involved for both processes, the similar effects are expected. It has been assumed that the capacity requirement figures for primary sedimentation apply for secondary clarification as well. The benefits are computed by considering capital, operating, and maintenance costs and the resultant savings are given below:

TABLE 7: COST SAVINGS FOR SECONDARY CLARIFICATION
(Heat Transfer Costs Excluded)

Cost Savings in Cents/1,000 Gal	Treatment Plant Capacity (MGD) 1	10	100
New plants	0.34 - 0.44	0.50 - 0.66	0.36 - 0.47
Existing plants	0.28 - 0.48	0.41 - 0.73	0.30 - 0.52

Disinfection — The two possible benefits of increased temperature operation on disinfection are lower chlorine requirements and smaller contact chambers. This is a tradeoff situation whereby for the same contact time less chlorine is required or at the same chlorine dosage rate less contact time is required. Each of these alternatives must be studied to see which offers the greatest advantage.

It was found that operation at a temperature level increase of 10°C can yield a savings in chlorine usage of from 39 to 66% or a savings in contact time of 31%. The chlorine dosage savings calculation is based on a normal input of 8 mg/l of chlorine recommended for an activated sludge plant. All of these calculations are for the indirect heat transfer method only since the use of direct heat transfer would result in additional dilution of the chlorine and more total chlorine would be required.

These benefits identified can be converted to cost figures by considering chlorine costs and the costs for equipment, operation, and maintenance in a disinfection facility. The savings for this process are given below:

TABLE 8: COST SAVINGS FOR DISINFECTION (Heat Transfer Costs Excluded)

Cost Savings in Cents/1,000 Gal	Treatment Plant Capacity (MGD) 1	10	100
Chlorine savings	0.20-0.33	0.20-0.33	0.20-0.33
Contact chamber savings	0.22	0.12	0.06

From this table it appears that the best approach is to take the chlorine savings. Since the contact chamber savings are competitive, at least for the lower capacity plants, a decision for the particular plant would rest upon the actual chlorine demand for the wastewater being treated and the design features of the disinfection facilities.

Activated Sludge Process – The temperature effects on activated sludge have been discussed previously. It appears that operation at higher temperature can yield higher removal efficiencies provided the higher temperature is not above the optimum of about 26°C. BOD removal increases from 82 to 91% when changing the operating temperature from 10° to 20°C.

As for sedimentation, this removal efficiency increase could be used to permit lower capacity units for new plants or to save in additional capacity which would otherwise have to be added to existing plants. The type of information needed to make calculations are:

(1) the reduced detention time (and corresponding capacity) possible in operating at the same BOD removal level, but at higher temperatures; and

(2) the increased detention time required in operating at the same temperature, but higher BOD removal rate. The results of such calculations for activated sludge are given in the table shown on the following page.

TABLE 9: EFFECTS OF WASTE HEAT UTILIZATION ON ACTIVATED SLUDGE

	New Plants	Existing Plants
Capacity required without waste heat use	1.00	1.30 - 1.37
Capacity required with use of waste heat (Indirect heat transfer: 10°C rise)	0.72 - 0.77	1.0

These benefits are quantified by a consideration of capital, operating, and maintenance costs. The resultant savings are given below.

TABLE 10: COST SAVINGS FOR ACTIVATED SLUDGE (Heat Transfer Costs not Included)

	Treatment Plant Capacity (MGD)		
Cost Savings in Cents/1,000 Gal	**1**	**10**	**100**
New plants	0.48 - 0.58	0.32 - 0.40	0.26 - 0.31
Existing plants	0.62 - 0.77	0.43 - 0.52	0.34 - 0.41

Nitrification — The temperature effects on nitrification have been previously discussed and were found to be intimately associated with the activated sludge process. Operation at higher temperatures gives higher process efficiencies. Table 11 indicates that at elevated temperatures lower aeration volumes and lower oxygen requirements can be used for nitrification. Based on this information, the following can be calculated:

(1) the reduced processing capacity possible in operating at a higher temperature, but yielding the same level of nitrification; and

(2) the increased capacity necessary to operate at the same temperature, but yielding a higher level of nitrification.

TABLE 11: EFFECTS OF WASTE HEAT UTILIZATION ON NITRIFICATION

	New Plants	Existing Plants
Capacity required without waste heat use	1.00	1.89
Capacity required with use of waste heat (Indirect heat transfer: 10°C rise)	0.53	1.00

In addition to the savings associated with unit capacities as outlined above, savings in compressor capital costs and compressor operating costs are possible due to the lower oxygen requirements. Both of these benefits are quantified by a consideration of capital, operating, and maintenance costs. The resultant savings are shown in Table 12.

TABLE 12: COST SAVINGS FOR NITRIFICATION*
(Heat Transfer Costs not Included)

Cost Savings in Cents/1,000 Gal	Treatment Plant Capacity (MGD)		
	1	10	100
New plants	0.97 - 0.98	0.65 - 0.67	0.52 - 0.53
Existing plants	1.8 - 2.4	1.2 - 1.3	0.82 - 1.0

*It is recognized that nitrification and the activated sludge process take place in the same equipment. Costs savings for waste heat utilization in activated sludge have already been calculated and the savings shown here are only the additional incremental savings in the system for the nitrification portion of the process.

Vacuum Filtration — Increased temperature has the effect of reducing the filtration area required in filtering sludges. An increase of 10°C gives an 11.6% reduction in filtration area. This results directly in reduced filtration equipment and reduced associated operating and maintenance costs. The savings for various treatment plant sizes are shown below.

TABLE 13: COST SAVINGS FOR VACUUM FILTRATION
(Heat Transfer Costs not Included)

	Treatment Plant Capacity (MGD)		
	1	10	100
Cost savings, cents/1,000 gal	0.32	0.31	0.23

Overall Cost Savings — If the waste heat is utilized in warming the wastewater before primary clarification, the benefits of operation at higher temperature are gained throughout the entire plant. Benefits for each of the processes can, therefore, be added together to obtain the overall cost savings. The resultant savings are given in Table 14 shown on the following page.

TABLE 14: OVERALL COST SAVINGS FOR SECONDARY TREATMENT PLANT (Heat Transfer Costs Excluded)

Cost Savings in Cents/1,000 Gal		Treatment Plant Capacity (MGD) 1	10	100
New plants	Without nitrification	1.54 - 1.90	1.52 - 1.95	1.27 - 1.61
Existing plants	Without nitrification	1.59 - 2.15	1.51 - 2.06	1.26 - 1.78
New plants	With nitrification	2.51 - 2.88	2.17 - 2.62	1.79 - 2.14
Existing plants	With nitrification	3.39 - 4.55	2.71 - 3.36	2.07 - 2.78

Benefit/Cost Analysis – The benefits of using waste heat were compared with the heat transfer costs derived previously. The first comparison was for the case where the waste heat is available directly at the treatment plant, such as for a combined utility or where heat transport involves no cost.

It was seen that benefits exceed costs for treatment plants with a capacity greater than 5 MGD. Depending on the design details of the particular plant, waste heat utilization may also be advantageous for plants with capacities between 2 and 5 MGD.

Where heat transport costs must be considered, the figures from Table 4 can be used. For this case, it appears that the benefits exceed costs for treatment plants of 5 MGD capacity or greater only when the waste heat must be transported for distances of less than about a half mile.

The above discussions are all applicable for a 10°C rise in wastewater temperature and for a final temperature of no more than 30°C. This temperature increase corresponds to the average which could be expected in using power plant cooling water. Areas where a higher temperature utility water were available are not so numerous, but the benefits from waste heat utilization in such areas are greater. It appears that if a 10°C rise could be obtained (and for final temperatures below 30°C), the following overall effects would apply:

(1) Benefits exceed costs for all size treatment plants where heat transport costs are negligible.

(2) For 10 MGD plants, waste heat utilization is economic for heat transport distances up to 2½ miles.

(3) For 100 MGD plants, it is economic for distances up to 4½ miles.

Another potential benefit of waste heat utilization in treatment plants is the reduction of the thermal shock mixing zone to receiving waters. Without utilization, the power plant hot water is discharged to the local receiving water at about 10°C above ambient.

With utilization, the power plant effluent can be used for heat transfer and then mixed with the treatment plant effluent to result in a combined discharge to the local receiving water at no more than 8°C above ambient (some heat losses will occur in the treatment plant, so this temperature will be somewhat lower in practice).

This benefit has not been quantified since a calculation of cost savings would depend on the thermal standards of the local area, heat losses in the treatment plant, and the method of heat dissipation being used by the power plant. Once again it must be indicated that the total heat load on the receiving water may be unchanged (or only very slightly lowered).

Temperature Effects on Advanced Waste Treatment Processes

In this section, each of the candidate advanced waste treatment processes will be examined to determine the performance characteristics at various temperature levels.

Reverse Osmosis – In reverse osmosis treatment, the wastewater is pumped under pressure into a membrane assembly. Purified water passes through the membrane and the concentrated brine (with impurities) is rejected by the membrane. The most important thermal effect on such a system is that operation above a reference temperature yields a higher product water flux rate through the membrane.

This means that increasing the temperature results in a smaller membrane area required to produce a specified produce water rate and quality. This effect is expressed quantitatively by saying that the required membrane area decreases 3.2% for every degree centigrade above a reference temperature for the same amount of product water.

Current reverse osmosis membranes are subject to performance degradation by hydrolysis. Hydrolysis increases with increasing temperature and 38°C is generally taken as the limiting operating temperature. Even below 38°C, the increased hydrolysis at elevated temperatures results in a poorer effluent and a shorter membrane life with corresponding increased replacement and maintenance costs.

Activated Carbon Adsorption – The subject of temperature effects on adsorption is very complex and the results of experiments in this field are somewhat confusing. It appears that the adsorption of some materials is increased with increasing temperature while other materials show decreased adsorption. In addition, the direction of the change in adsorption for some materials with temperature change depends on the type of activated carbon utilized.

Detailed studies with adsorbing biochemically resistant materials provide an insight into this confusion. Experimental work shows that adsorption rate increases with increasing temperature. The time to reach equilibrium is a variable, depending on the material being adsorbed, and may range from weeks to months.

Other experimental work shows that adsorption at lower temperatures tends to approach equilibrium more slowly and the equilibrium value of adsorption capacity is greater. In summary, adsorption at a temperature above a reference level takes place more rapidly and approaches equilibrium sooner, although the equilibrium value of adsorption capacity is lower.

In an actual treatment system, true equilibrium may never be reached and small differences in equilibrium adsorption capacity, therefore, may be of no importance. Another possibility is that the lower capacity at elevated temperature operation (which could require more frequent regeneration of the carbon) may be more than compensated for by the higher quality effluent resulting from the higher adsorption rates.

An examination of experimental data in the literature reveals that in the range from about 10° to 30°C, an increase in temperature of 10°C gives an 8% reduction in equilibrium adsorption capacity, but a 36% increase in initial adsorption rate.

Ion Exchange – Temperature effects on ion exchange processes are also very complex. Whereas the dependence of adsorption on temperatures has not yet been studied systematically, theory indicates that highly specific adsorption decreases with increasing temperature. The effect of temperature on exchange selectivity seems to depend on the specific ion involved.

Some ions, such as sodium, potassium, and barium, have decreased selectivity at higher temperatures, while other ions, such as zinc, cobalt, and beryllium, have increased selectivity. Ion exchange rates increase with increasing temperatures, depending on whether particle diffusion control applies (where a 4 to 8% increase per degree centigrade is possible) or film diffusion control (3 to 5% increase per degree centigrade). Elevated temperatures may be impractical for use if temperature sensitive ion exchange resins and/or chemicals (e.g., ammonia) are utilized.

Ammonia Stripping – Ammonia stripping basically involves blowing air through wastewater to transfer dissolved ammonia from the water into the gas phase. Raising the temperature of the wastewater has the effect of making the ammonia less soluble and more easily removed. This reduction in the solubility of ammonia vs temperature is indicated in Table 15.

TABLE 15: SOLUBILITY OF AMMONIA IN WATER

Temperature, °C	Solubility (Parts Ammonia per 100 Parts Water
0	88
10	69
20	53
30	41
40	32
50	23

The concentration of ammonia in typical wastewaters, however, is on the order of 25 mg/l (or 0.0025 parts/100 parts water) and the solubility reductions of the above table would be of little use in increasing stripping efficiency.

Increasing wastewater temperature also has the effect of increasing the partial pressure of ammonia which promotes removal. An increase of 10°C in the range of 0° to 50°C results in an increased ammonia partial pressure of from 59 to 80%.

A stripping operation at elevated temperature should, therefore, be capable of a higher rate of ammonia removal for the same air flow rate or the same ammonia removal for a lower air flow rate (assuming the stripping air can effectively remove the higher quantities of ammonia in the vapor phase due to increased vapor pressure). Since ammonia stripping does involve blowing ambient air through the wastewater, this process is vulnerable to problems of freezing during cold weather. Heating the wastewater would permit operation of the stripping unit during periods when the weather would normally force the unit to be inoperative.

Heating the stripping air could also be done in this case, but this is a more expensive method (especially if the heated wastewater was already used in previous operations). Use of heated air or water would have the added advantage of reducing possible calcium carbonate fouling in winter.

Biological Denitrification – This is basically an anaerobic process in which the nitrate ion is metabolized by microorganisms to nitrogen gas. Since it is a biological process, it is expected that increasing temperature will increase metabolic action and denitrification.

Phosphorus Removal by Precipitation – This process involves the addition of chemicals to wastewater which react with the phosphorus to yield insoluble phosphates. A wide variety of chemicals have been tried (such as lime, iron salts, aluminum salts, and lanthanum) and the point of chemical addition can be anywhere in the treatment train (primary, secondary, or tertiary).

Elevated temperatures could be beneficial in phosphorus removal by increasing the chemical reaction rate or by shifting the equilibrium to favor increased phosphorus removal. A study of temperature effects using aluminum and iron salts shows, however, that reaction temperature appears to have no effect on the rate or extent of phosphorus removal.

Elevated temperatures could also be useful in speeding the sedimentation of the precipitates for this process just as for suspended solids in normal sedimentation. This benefit of higher temperature will not have much effect on overall phosphorus removal efficiencies since systems are currently designed to operate at 85 to 95% removal, but the more rapid sedimentation will reduce required detention times which means that smaller sedimentation tanks can be used.

Assuming that the precipitates in phosphorus removal respond in a similar manner to suspended solids in conventional wastewater treatment sedimentation, sedimentation tanks in a phosphorus removal system could be 15 to 25% smaller for a 10°C increase in temperature.

Heat Transfer Benefits for Advanced Waste Treatment Processes

As for the normal biological treatment processes, benefits for advanced treatment processes due to waste heat application are expressed as savings in cents/1,000 gal treated. Each of the processes will be examined in the following paragraphs to determine benefits on an individual basis. The overall benefits for a specific plant will depend on which processes are incorporated into the plant and which ones have the increased temperature.

Reverse Osmosis – Required membrane area for a given capacity is reduced 3.2% for every degree centigrade increase as shown previously. For the 10°C rise taken as a standard for this study, the net membrane reduction is 32.4%. With this benefit value, the savings in capital, operating and maintenance costs for reverse osmosis can be calculated. The results are given below.

TABLE 16: BENEFITS FOR REVERSE OSMOSIS
(Heat Transfer Costs Excluded)

	Treatment Plant Capacity (MGD)		
	1	10	50*
Savings in cents /1,000 gal	9.4	7.3	6.1

*This is the maximum size reverse osmosis plant for which costs have been estimated.

Heat transfer costs are shown in Table 4. A comparison of benefits and costs shows that waste heat utilization is economical for all size treatment plants using reverse osmosis where there are no heat transport costs (as for a combined utility complex). Where the waste heat must be transported from a distance, utilization is still economical under the following conditions:

(1) For 1 MGD plants where the distance is no more than 3½ miles.
(2) For 10 MGD plants and distances up to 14 miles.
(3) For 50 MGD plants and distances up to 13 miles.

It must be remembered that elevated temperature operation of reverse osmosis also yields a somewhat poorer effluent and lower membrane life.

Activated Carbon Adsorption – As discussed previously, a 10°C increase in wastewater temperature was seen to result in a 36% increase in adsorption rate but an 8% reduction in adsorption capacity. It can be assumed that the increased adsorption rate permits a smaller capacity adsorption unit to achieve the same effluent quality. The lower adsorption capacity will require more frequent regeneration. These two conflicting factors must be evaluated in terms of capital, operating, and maintenance costs to determine the possible benefits of waste heat utilization. The results of this evaluation are given below.

TABLE 17: BENEFITS FOR CARBON ADSORPTION (Heat Transfer Costs not Included)

	Treatment Plant Capacity (MGD)		
	1	10	100
Savings in cents/1,000 gal	3.60	1.80	0.86

A comparison of these savings with heat transfer costs reveals that benefits exceed costs for waste heat use in carbon adsorption only for plants with capacities less than about 60 MGD and where there are no heat transport costs. Where heat transport costs must be considered, plants in the range of from 1 to 10 MGD capacity can economically utilize waste heat only where distances are less than about one-half mile.

Ion Exchange – Ion exchange rates increase from 3 to 8% per degree centigrade with increasing temperature, as previously indicated. The assumed standard of 10°C rise yields increased rates of from 30 to 80%. The rates can be equated to smaller exchange units to achieve the same quality effluent. By considering costs for ion exchange, savings resulting from waste heat utilization can be calculated. The results of these calculations are given below.

TABLE 18: BENEFITS FOR ION EXCHANGE (Heat Transfer Costs Excluded)

	Treatment Plant Capacity (MGD)	
	1	10
Savings in cents/1,000 gal	1.6 - 3.1	0.8 - 1.5

A comparison of benefits and costs indicates that waste heat utilization could be economical for ion exchange units of from 1 to 10 MGD capacity and where the waste heat must be transported no more than about one-quarter mile. These savings would probably represent upper limits since temperature effects for ion exchange processes are not fully defined.

Ammonia Stripping – The 10°C wastewater temperature increase appeared to increase ammonia removal efficiency from 50 to 80% as discussed earlier. Results of actual plant operation at different temperatures show that to yield the same quality effluent, the 10°C rise permits the use of small stripping units (from 53 to 67% of normal size). With this information and a review of capital, operating, and maintenance costs for ammonia stripping benefits of waste heat utilization can be determined. The resultant savings are given below.

TABLE 19: BENEFITS FOR AMMONIA STRIPPING (Excluding Heat Transfer Costs)

	Treatment Plant Capacity (MGD)		
	1	10	100
Savings in cents/1,000 gal	1.1 - 1.5	0.5 - 0.7	0.5 - 0.7

A comparison of these savings with costs reveals that waste heat utilization for ammonia stripping alone is not an economical process. Where ammonia stripping is used in conjunction with other processes, however, the savings of stripping can be added to the savings from the other processes.

Biological Denitrification – The denitrification rate is increased 12% per degree centigrade temperature increase, as shown earlier. This will yield a 120% rate increase for the standard 10°C rise. On this basis, the denitrification chamber capacity can be reduced to about 45% of normal size and still produce the same quality effluent.

Costing information on denitrification is not abundant, but a recent report indicates that the capital cost of such a system would be about 25% greater than for a conventional activated sludge process of the same flow. The denitrification chamber accounts for only about 20% of the total in this overall nitrification/denitrification system. A consideration of activated sludge process costs permits an estimation of the benefits for denitrification due to waste heat utilization as shown on the following page.

TABLE 20: BENEFITS FOR DENITRIFICATION
(Heat Transfer Costs not Included)

	Treatment Plant Size (MGD)		
	1	**10**	**100**
Savings in cents/1,000 gal	0.12	0.08	0.07

Comparing benefits and costs reveals that waste heat use for denitrification alone is not economical. Since this process is used in combination with others, however, the benefits can be added to those of the other processes.

Phosphorus Removal by Precipitation — It was previously seen that sedimentation tanks in a phosphorus removal system could be 15 to 25% smaller for a 10°C increase. With this range of benefits and a consideration of precipitation costs in a two-clarifier process, savings can be calculated. The results are:

TABLE 21: BENEFITS FOR PHOSPHORUS REMOVAL
(Heat Transfer Costs not Included)

	Treatment Plant Capacity (MGD)		
	1	**10**	**100**
Savings in cents/1,000 gal	0.76 - 1.13	0.49 - 0.70	0.37 - 0.52

Comparing benefits and costs shows that phosphorus removal alone does not show promise in profiting from waste heat utilization, but the savings of this process should be considered in a system combining many processes.

Benefit/Cost Analysis — The benefits derived above can now be compared with the heat transfer costs. The first comparison is for the case of a combined utility where heat transport costs are negligible and where the process is considered by itself. It appears that benefits exceed costs for all sized reverse osmosis plants, for carbon adsorption plants up to 60 MGD, and for ion exchange plants up to 10 MGD.

Costs exceed benefits for ammonia stripping, denitrification, and phosphate removal. The second comparison is for the case of heat transport distances of a mile or more, and again, where the process is considered by itself. In this case reverse osmosis alone is favorable and for distances of up to 14 miles of heat transport. The final comparison is for the case of the process considered as an adjunct to a secondary treatment plant. In this case, all processes are favorable since positive benefits are apparent for each and the heat transfer costs have already been charged to the secondary plant (no additional costs are required if the effluent from the secondary plant is already heated).

Research Needs

The knowledge of present and projected environmental efforts resulting from large injections of waste heat into the environment by means of wet and dry cooling towers and cooling ponds, lakes and streams, is extremely limited and requires extensive research.

A conference was held in Zion, Illinois in September 1971 sponsored by the National Science Foundation to estimate research needs on problems of waste heat transfer from large thermal sources namely electric generating plants. W.C. Ackermann of the Illinois State Water Survey prepared a report of this which was published as PB 219,056. The material in this chapter is based on that report.

BACKGROUND

In recent years, the use of large cooling towers with direct heat dissipation to the atmosphere has increased, particularly in Western Europe, and several such installations are now in operation also in the United States. Very large cooling towers in arrays, either of mechanical draft or natural draft design, and large arrays of cooling ponds represent much more rapid means of releasing heat and moisture to the atmosphere than the release through large natural water bodies.

However, the direct atmospheric release of large amounts of heat and moisture immediately raises questions as to the further ecological and meteorological consequences.

Essentially the lack of knowledge of atmospheric effects from large cooling towers, ponds or reservoirs has not permitted a conclusive scientific decision on the least detrimental means of heat dissipation for new power plants. The ecological effects have never been seriously considered except in streams. Near oceanic areas, the fallout of salt from the use of ocean water in cooling towers may be a problem.

Investigations conducted by industry and governmental agencies have all indicated that atmospheric changes may occur in general, although the type of conditions altered and the magnitude of effects are uncertain. All agree that there may be some increased fog and icing, but the magnitude and the significance of these effects are either not stated or are debatable.

The possibility of increased atmospheric turbulence, growth of clouds, production of rainfall and snowfall, severe local storms, and alterations in other weather conditions cannot yet be assessed adequately from mathematical models or observational data. Observations and studies have shown that observed plumes seed clouds and that rainfall increases are measurable, but detailed supporting data are not available.

The effects from plumes will be greatest at night, the most difficult time to measure and record. Thus, climatological studies and application of meteorological theory to the problem have been inconclusive and have led to differences in opinion as to atmospheric effects from cooling towers.

Large cooling ponds or channels with spray capabilities have recently been introduced in the Midwest after several years of study in the southern states. The effects from increased invisible moisture in the atmosphere on the social and ecological aspects in the immediate area are not known.

Icing on highways is a problem in Minnesota and Wisconsin. The design of power generation plants offshore is also being considered since the heat dissipation at sea is not considered to have any immediate major effects on the environment. However, little is really known about the consequences of these new heat dissipation methods.

RESEARCH PRIORITY AREAS

For clarity and simplicity in presentation of the waste heat problem it was decided to describe some of the problems and the proposed research by the various disciplines involved. Hence, the background and research needs are presented in this order, social, biological, atmospheric, hydrologic, and technology/engineering.

Social

From a sociological perspective, the proposed application of new methods (other than once-through cooling) for handling waste heat from electric generation plants may be seen as the application of a new technology which may produce physical effects in the atmosphere and in turn have a variety of social and economic impacts. This proposed application, therefore, has many similarities to planned weather modification efforts.

Relationship to Other Societal Problems: Research results from weather modification studies suggest that the following variables may be relevant in understanding the social response to the proposed use of large cooling towers, lakes, or similar applications.

Views of the more powerful or more vocal interest groups and opinion leaders toward the agencies or industries, known to be in support of or proposing to utilize the new technology – Enemies of the power industry are likely to become active; friends may or may not get involved. Credibility is a key element here.

The financial resources and the legal and political sophistication of such interest groups and opinion leaders – Any long term, effective action requires readily available resources and skills.

Anticipated economic impact that would follow the installation of the new technology – Those anticipating significant economic gain may or may not take action, but those believing that they will suffer economic loss are most likely to make their voices heard.

Level of consensus among recognized experts regarding the probable physical effects of the new technology – Where the experts don't agree, every would-be protagonist can pick his preferred expert and have a fair chance of winning converts to his point of view. Thus, public hearings may have the logical character of an argument among a group of children.

The decision makers' perception of public opinion regarding the proposed application – The real public opinion is seldom known. The decision makers' perception of it will be shaped by the mass media, organized letter and phone campaigns, and the individual decision maker's selective perception from his personal encounters.

Mobility of activist groups – Where organized opposition has emerged in one area, a proposed installation elsewhere may serve as an attraction. On occasion it really is true that it is the outsiders who have stirred up the trouble.

Anticipated aesthetic degradation – The proposed plant with its new technology may be seen as a potential visual blight in and of itself as well as a destruction of otherwise desirable open space or green area.

These are just some of the social factors which may come into play prior to the building of the installation. Once the plant with its waste heat dissipation technology has begun to operate, other considerations may come to the fore. Examples would include the following possibilities.

Concern about humidity induced problems such as respiratory problems, paint deterioration and the corrosion of metals, increase in fungus diseases, and personal discomfort attributed to high humidity levels.

Concern produced by the high visibility of plumes and fog. Plumes from nuclear powered plants may be viewed as hazardous.

Concern about economic impact such as a negative effect on land values and retail sales attributed to the reputation of near the plant and downwind locations.

Concern about aesthetic degradation.

Concern about accidents purportedly resulting from plume induced fog, icing, rain, or snow.

Research Needs: Needs in the social area involve two different major areas of research. The research needs related to the matters recounted above can perhaps be best phrased as follows. What procedures and emphasis can be used to provide an adequate basis for the development of informed consent or informed rejection in communities where new electric generation plants are being proposed?

Since there appear to be many interrelated factors potentially involved in any worthwhile informed consent process, and given the dearth of research on the social implications of new electric generation plants, the most useful research approach initially would be to conduct a series of intensive case studies in communities where the new technology has already been applied and where such installations will be constructed in the next few years.

Where possible, matching communities could be selected where one has once-through cooling, one a cooling tower, and one a man-made lake. Where feasible, careful surveys should be made of initial public opinion and views held by powerful and vocal interest groups. Then the informing and decision processes should be followed in detail. The monitoring of social response to the installation and operation of the plant should be continued for six months to a year.

These case studies should be preceded by observation and informal data collection in various areas in Western Europe where cooling towers have been used extensively for some years. In the United States there are only about 15 sites where large cooling towers are in use and most of these have been installed in recent years.

Specific questions to be addressed in the case studies would include: What are the issues that most typically arise? How is information and misinformation relevant to these issues secured and utilized? What are the factors which appear to produce controversy and court suits? What procedures appear to offer the greatest hope for rational decision making and the fair adjudication of grievances?

A second major area of research seems warranted. We need to know to what extent and in what ways is it possible to bring about a voluntary reduction in the approximately 7% per year increase in the peak consumption of electric power.

This question centers, in large measure, on the basic research question of how daily habit patterns or life style can be altered without the threat or application of legal action. The answers to that issue will apply not only to the use of electric power but also to a whole range of environmental concerns.

Here some creative, experimental, nonlaboratory pilot studies appear to be in order. A minimum period of two years will be required to get a first approximate answer to the question. This should be collaborative research involving sociologists, economists, and social psychologists. Both types of research envisioned here, to be successful, will require the active cooperation of the relevant electric power companies.

Biological

Waste heat additions to the terrestrial or aquatic environment will exert an influence on the ecosystem. However, it is generally recognized that the direction, the magnitude, and the significance of the effect are not understood nor can they be fully defined for any particular ecological complex.

As a consequence of numerous biological phenomena and their responses to environmental variables there is need for comprehensive data and information at all biotic levels as well as on critical members or species of the ecosystem.

Environmental influences, including waste heat increments (in amount or extra duration of exposure), can be expected to elicit changes which may be small and of little populational or species significance, or which may be of great magnitude and marked biological effect.

However, these effects must be studied from the purview that they are, or will be, superimposed upon a background of natural population and environmental oscillations in aquatic environments which are of diverse quality, and which are, in addition, modified by regional variations. Since these variations correlate with both short and long time scales, accurate assessment of any ecosystem will correspondingly require study on comparable scales.

Diversity of the natural environment requires a distinction between the major categories of terrestrial and aquatic, but the latter is further diversified into streams of variable flow velocities and volumes, lake (natural and artificial), estuaries, and other coastal marine waters having variable physical and hydrographic characteristics. These environments each provide the matrix for substantially different ecosystems. Latitudinal and other geographic differences impose a further need for regional studies.

Comprehensive understanding of the aquatic ecosystem can be accelerated by study of existing areas of thermal modification, including river, lake, and estuarine waters. Data obtained through laboratory simulations provide useful information, but it is generally understood that such studies yield supplemental data only and cannot substitute for the profoundly more complex field conditions. It is unrealistic to attempt to evaluate biological effects by utilizing a single environmental variable such as temperature. While multivariate studies are designed to duplicate environmental conditions, this has not been done because of lack of complete information on each environment. Laboratory studies will allow measurement of critical thermal limits and studies of age-class thermal sensitivity and its influence on the behavior of index species, as well as other biological effects resulting from elevated temperatures.

Such studies may provide insight into the range of responses that will be elicited under natural conditions and thereby increase the efficiency of in situ environmental research. However, the results obtained must be interpreted within the context of natural environmental characteristics, oscillations, and cycles.

Terrestrial Environments: Mechanisms of transfer of waste heat discharges from large sources into the atmospheric environment include cooling towers of diverse design, lakes, ponds, and sprays. The concerns for ecological effects, resultant from thermal additions, are primarily those beyond the boundaries of the land occupied by the power complex. Significant biological changes may occur within those boundaries as a result of dispersing heat from a given installation.

However, loss of productive land for crops, forestry, wildlife, and recreation, among other uses, to accommodate the construction of cooling towers, lakes, or multiple sprays is, in principle, no different from any major land conversion from one use to another. Ecological concern is directed primarily toward external effects, such as the possibility that the installation would stimulate reproduction of harmful insects or plant species which would disperse into neighboring areas.

It is believed that present knowledge is generally adequate for a preliminary assessment of the probable ecological consequences of heat discharge to the environment of terrestrial plants and animals. Because each site, and each anticipated means of heat dissipation, is to some degree unique, an assessment of probable ecological effects should be made as a part of the site selection process and system design.

It is estimated that the effects of heat discharge (including those of associated humidity) on terrestrial organisms will be principally: changes in timing and rapidity of growth of fungal diseases of plants and animals (mostly insects); changes in evaporation and transpiration due to decreased solar radiation (fog) and increased humidity;

interference with wildlife migration routes or breeding sites; influence of warmer temperatures at the margins of cooling ponds, lakes, rivers, or other heat disposal sites on the habitat or refuge of desirable or pest species or communities; and changes in precipitation. The research needs identified in the above fields are as follows:

> Evaluate jointly, the extent and duration of humidity changes and their relation to temperature cycles in the vicinity of planned facilities. As meteorologists and engineers estimate the statistical probability of an increase in ground level humidity, temperature, or precipitation for a particular cooling facility, agronomists, plant, and insect physiologists and ecologists should then judge the influence these changes may have on
>
> the growing season
>
> the influence of aerial changes including fog and solar radiation, on transpiration and evaporation of the principal flora or plant communities. Existing computer models designed to assess transpiration and evaporation on seasonally and regionally based soil moisture and plant growth should be coupled with meteorological and engineering predictions relative to cooling facilities.
>
> the influence these changes may exert on pest, pathogenic, and beneficial organisms which are endemic to the area of the planned cooling facility (e.g., spore germination, fungal growth, insect life cycles, etc.).
>
> Evaluate the influence that these changes may have on areas adjacent to those planned for the location of cooling towers or other heat transfer installations. Evaluations should include the effect on animal refuges or habitats and on useful and pest species whose life cycles might be altered by these local changes.
>
> Evaluate the influence of cooling towers which have been in operation both in this country and others (e.g., Western Europe) through comprehensive field studies. These should include a comparison of essentially identical or equivalent areas with respect to plant and animal communities but which are beyond the immediate influence of the cooling facility.

Aquatic Environments: Concern for the environmental effects of heat transfer to water resources is directly related to the extent of the area involved and the rate of heat dissipation. The geometry of the thermal plume and its gradient, from the source (outfall) to a point of negligible variance from ambient conditions, defines the area of direct influence as well as the magnitude of heat transfer to the aquatic ecosystem.

Modeling of thermal plumes is essential to predict the area, volume, and temperatures for proposed cooling water discharges, and to define the minimum area which must be subject to intensive biological investigation.

Rivers and Streams – Generally accepted and legitimate uses of rivers include navigation, recreation, water supply, hydropower generation, wastewater assimilation, agricultural consumption, and the maintenance and support of biotic communities.

Regionally there may be varying priorities of one use over another resulting from riparian rights, local developments, and social and economic needs of the area comprising the river system. Each of these uses, in some way, influences the water quality and may result in a beneficial or detrimental effect.

For example, an increase in water temperature could be beneficial to navigation during ice periods; alternately, decreased visibility due to fog formation could exert a limitation on this water use. Furthermore, increased temperature accelerates biochemical oxidation, in addition, it develops an added oxygen deficit. This thermal effect could result in a local decrease in water quality, but a more rapid recovery of a deteriorated river system over a greater distance, and ultimately serve to enhance its water quality.

Thus, the interaction of increased thermal levels with other water characteristics must be elucidated within the scope of multiple uses and water quality specifications. The complementary and antagonistic effects consequent to thermal additions to a waterway serving multiple and diverse needs must be understood. Studies to achieve this goal must be regional since hydrographic and biological characteristics of a river system, and the relative priorities of its use, vary on a regional basis.

Comprehensive studies are needed to achieve a balance among various water uses and its conservation. The specification of a thermal standard would result from careful consideration of all constraints imposed by present and future uses in terms of the impact of a thermal increment on its other legitimate uses. Such understanding is a necessary prologue to resolving the questions as to the best environmental route for a thermal discharge, e.g., into a waterway or to the atmosphere. Based upon the differences in river characteristics and multiple uses, study should be directed as follows:

> Investigate the biotic, chemical and physical effects of thermal additions at the immediate point of discharge of warmed water and along the gradient of these effects for the distance influenced.
>
> > Clarify the thermal effects on biotic populations as they vary with season and volume flow; these must include annual extremes of climatic conditions.
> >
> > Determine whether thermally induced changes in the ecosystem are reversible, and the time scale for recolonization as well as the nature and diversity of newly established populations.
> >
> > Analyze comparatively the relative intensity of the impact of thermal addition to a river system when contrasted with the impact resultant from use of alternative methods of heat transfer.
>
> Through long term studies, elucidate the extent and duration of the impact of thermal addition to the river on water and air temperatures with respect to shifts in the annual averages and their range.
>
> Study the impact of thermal addition to a river system to elucidate its interrelation with other uses of these resources. Such studies should clarify priorities related to social, economic, and conservation needs of the area which depends upon the river system.

It is noted also that specific biological studies indicated below as research needs for lakes, estuaries, and marine waters are applicable to study of rivers and streams.

Freshwater Lakes, Estuaries, and Marine Waters – Ecosystem responses to thermal increments will be different according to the characteristics of the environment being considered for use in energy production and waste heat transfer. An accurate assessment of change can be made only with a background of factual information which includes a comprehensive structural and functional description of the ecosystem in question.

Within this concept, collection of basic data must be continued for each category of aquatic environment to provide sufficient information for the development of reliable models essential for predictions and planning. After basic data have been acquired for each category of aquatic ecosystem, the consequences of environmental modification can be more clearly identified.

If with continued engineering development there is a decrease in the thermal discharge into aquatic resources, there will result a smaller modification of the environment. The nature of the biological response may become negligible, or the response may be that of very subtle long term changes which would be evident only through long term detailed study. The research needs are identified as follows:

Study the interaction between temperature and reactivity of chemical constituents of the aquatic environment. These studies would be essential principally in areas unique with respect to chemical composition, e.g., high salinity. In addition, thermal effects on buffering capacity, chemical reactions, and rates of chemical change should be studied as they relate to thermal alteration of the environment. These needs pertain predominantly to estuarine and marine waters; to areas where effluent brines will be both heated and concentrated; and to areas where chemical pollutants already exist, or may be anticipated to increase.

Study community metabolism to provide initial evidence on the impact of thermal additions to each type of aquatic ecosystem. The relationship between gross productivity and respiration provides a basis for estimating biological change with respect to eutrophication, to increased production, or to a breakdown in energy flow through trophic levels. Information in this category is more quickly obtained and would provide a more rapid assessment of ecosystem modification than other more detailed (but necessary) and time consuming studies.

Identify key or index species at the significant trophic levels and species of sport and commercial value. Variance in response of these species to temperature as a function of age, sex, and past physiological and thermal history must be determined. Essential information must be acquired for:

Thermal tolerance limits and preferences per se and with variable temperature histories.

Fecundity and the influence of chronic thermal

increments on it.

Behavioral responses to changing thermal regimes; these should include social (grouping) behavior, migrations (diurnal and seasonal, vertical and horizontal), predator-prey relations, and parasite-host relations.

Magnitude of energy turnover at both natural and increased acclimation thermal levels.

Study long term effects of small temperature increments on interspecies competition (expressed in shifting diversity values).

Study selection by fishes of new positions in thermal gradients and its correlation to food dependence and availability.

Determine the effect of entrainment of planktonic organisms through cooling systems and its significance on the area ecosystem with respect to trophic relations and species diversity.

Determine the effect of blow down on the immediate area ecosystem, with particular reference to the interactions of these concentrated chemicals and temperature.

Study the influence of thermally based shifts in oxygen saturation levels on metabolism of organisms permanently resident (benthos and periphyton) in the area of thermal discharge.

Determine the significance of the impact of thermal increments on each of these aquatic resources. For this a comprehensive study of the ecosystem is essential and should include:

Primary productivity and biomass.

Species diversity and population density at each trophic level.

Energy transfer through trophic levels.

The nature of the chemical and physical matrix of the system.

The variations in this system concurrent with seasonal and annual cycles.

Population (species) grouping within and at the interface of (thermal) mixing zones.

The climatological overlay as it relates to surface water movements, upwelling, turnovers, current velocity and direction, temperature modifications, and radiant energy availability should be correlated with structural and functional aspects of the described ecosystem.

With the development of artificial lakes, ponds, and impoundments, study their ecological history expressed in the establishment of individual populations and communities. Such information is not available and it would complement populational studies in natural lakes. These studies would provide a base for predictions concerning future development of artificial lakes for waste heat transfer purposes.

Deep Ocean Water – It has been suggested that nuclear power complexes or their heat discharge facilities could be constructed off shore in the ocean. Buoyancy differences consequent to reactor cooling would result in deep water entrainment and vertical overturn of deeper nutrient-rich water inducing artificial upwelling which could increase surface fishery yields.

This route of heat transfer would minimize the temperature increase at the sea surface and would enrich this layer to support increased biological productivity. It has been estimated that oceanic areas of natural upwelling represent approximately 1% of the total ocean area and produce 50% of the fishery production. Some studies are in progress on the physical and biological characteristics of natural and artificial upwelling using sewage outfalls as an analog for nutrient upwelling. The research needs are identified as follows:

Conduct economic feasibility and reliability studies of cooling water outlet structures to provide maximum quantities of nutrient-rich deep water at the surface.

Develop models of the quantities of deep ocean water which would be brought to the surface. The spatial and temporal distribution of warmed waters resulting from each feasible combination of nuclear plant design, discharge configuration, and ocean current patterns must be defined.

Continue to develop and refine physical and biological models, supported by field and simulation study data, to estimate the consequences, and the beneficial use, of deep ocean heat transfer.

Atmospheric

The dissipation of waste heat from industrial processing either directly or indirectly occurs in the atmosphere and becomes part of the radiation budget of the planet. The future gigantic demands for power will create great amounts of waste heat for disposal in the atmosphere.

Most meteorologic and hydrologic observational programs are not designed to provide the required data to explain the consequences of the transfer of heat and moisture to the air environment. The following material describes some of the needs for baseline surveys, field observations, laboratory, and theoretical studies.

It will be necessary to collect specialized measurements and analyze data of a nonstandard nature in order to understand the impact of the various cooling systems on the environment. These data requirements, which are the largest felt need, are discussed in the following sections, grouped according to the heat discharge system. Studies for site selection require a different type of data from that required for research to understand the tower effects on the environment.

Natural and Mechanical Draft Wet Cooling Towers: Site and Preoperational Surveys – A climatological survey for a period of at least one year should be made prior to site selection to establish its suitability for heat and moisture discharges and dispersion. Local topographic features (hills, valleys, large lakes, etc.) may produce unacceptable environmental conditions. Thus, the climatological surveys should be tailored for each site. Items to be measured routinely and recorded every one or two hours include

Air temperature to nearest 1°C
Moisture content (saturation deficit would be better)
Wind speed and direction

In addition, the frequency and extent of fog, freezing fog, and dew should also be measured for several years before the power plant becomes operational.

Postoperational Data Needs – The data collecting program should continue and be expanded to include the geometrical and physical properties of the visible cooling tower plume. Photographic techniques can be used to measure the geometrical parameters to be correlated with plant, engineering, and meteorological data.

The invisible water vapor plume should also be monitored. Radar and weather satellite data can be used to locate areas of precipitation and cloud buildup to see if naturally occurring clouds and rainfall and subsequent mesoscale weather patterns are augmented by the cooling tower effluent. The surface fog network should be expanded to detect any increase due to the tower plume, remembering that the point of fog produced by the tower may be some distance from the source.

A network of temperature and humidity sensors (with accuracies better than those of hair hygrometers) should be used to detect surface temperature and humidity increases due to heat and water vapor being diffused to ground level. It is expected that these changes will be quite small and well within the natural variations. Documentation of this fact will be needed for regulatory activities.

In addition to these daily observations, intensive observations should be made for short periods (a few hours or days) during weather situations representative of all seasons and hours for model verification. These data should include:

Temperature, humidity, and wind profiles of 1 kilometer for distances up to 10 kilometers upwind and downwind of the cooling towers at the surface.

Temperature, liquid water content, drop size, and flow rate cross-sections at the tower opening and within the visible plume.

Cross-sections of temperature and water vapor in the invisible plume as far downwind as it can be identified.

Water sample collections to detect drift.

Nucleating properties of the drift particles.

Mechanical draft towers, because of their lower level of release and the nature of their discharge, are known to have a much greater potential than natural draft units to increase local ground level fog and humidity. Also, these meteorological changes will occur closer to mechanical units.

Dry Cooling Towers: The heat discharge from such units could trigger an existing instability. Radar techniques should be used to detect changes in clouds and induced precipitation.

Small Cooling Lakes and Ponds: These small lakes are defined as those with less than 1 to 3 acres of surface per megawatt of electrical energy produced. The major local changes to be expected will be in the intensity, frequency, and vertical and horizontal extent of fog, and the creation of freezing fog very near the water's edge. The primary concern to man will be the extent of the induced fog. A large number of expensive optical devices (transmissometers) may be needed to monitor conditions over a large area.

Fog and dew conditions, especially frequency and duration, should be measured at the site before the pond is built, after the pond is built but before its use for a heat sink, and during its use for cooling.

Mesoscale weather effects will be much more difficult to measure and monitor. Radar and weather satellite pictures could be used to detect whether cumulus clouds are generated by large cooling lakes.

Air crossing the cooling lake will be warmer, more humid, and less stable than it would otherwise be. These changes may be small and difficult to measure beyond a short distance from the pond's edge. A program to measure these changes should be initiated, if only to show how small the effect is.

Spray Ponds and Canals: Spray units greatly increase the effective area of evaporating surfaces. They will also increase the frequency and intensity of dew, fog, frost, and icing conditions along the banks or downwind of the canal or lake.

Since spray canals and lakes are very new, their impact on the environment has not been studied. Drift of droplets and invisible humidity may be a problem near the area of the sprays, especially in subfreezing weather. Consequently, the same remarks made for cooling lakes apply to spray ponds and canals.

Once-Through Cooling, Large Lake or Ocean: Because of mixing, a large fraction of the total heat from power plants will be added to the main body of water and later released slowly to the atmosphere, over a large area. Hence, local weather changes will be minor. An increase in local fogginess at the outfall may be detected.

A major research need is to determine the relative magnitude of the two processes in the water which lower plume temperature: turbulent mixing versus atmospheric losses which occur through evaporation, conduction, and radiation for a wide variety of weather conditions and for the various ways of releasing the heated water to the main water body (surface flow, submerged jets, etc.). This matter is described in more detail in the Hydrologic section.

Model Development—Cooling Towers: There is a need for sophisticated mathematical models that can be applied for given sites and meteorological conditions to define the behavior and effects of thermal discharge from cooling towers. Some models currently exist and are in use, however, these models require verification, further development and generalization. Some plume characteristics and effects have not been modeled to date.

Plumes from Natural Draft Cooling Towers – A model should be able to predict the ascent of the plume, its dimensions as a function of height, and its physical characteristics (temperature, water content, vertical velocity). Models based on cumulus cloud models are being applied to cooling tower plumes with apparently realistic results. Wider evaluation of the capability of these models is needed. In particular, certain parameterizations in the models should receive attention.

Both basic cloud physical processes (droplet growth, coalescence, freezing) and the mixing of plume and environmental air (entrainment) are normally handled by empirical parameters taken from cloud modeling experience. Even in the cloud models, parameter values and their dependence on cloud-type, geographical region, etc. have not been well established. Examination of the applicability of these parameterizations to cooling tower plume models is needed.

Less advanced than the modeling of natural draft plume rise is the capability for predicting downwind plume behavior and vertical diffusion. This element is critical because it determines when, and to what extent, moisture from a wet plume will exist at ground level. Thus, the determination of ground fog potential requires accurate specification of vertical diffusion from elevated sources.

Present models and limited field experience are in basic agreement that plumes from tall natural draft towers seldom cause ground level fog. But the conditions under which fog can occur and their relative frequency are not understood. Models should be developed and tested for specification of ground level concentration from plumes aloft, especially in inversion conditions. Because of the complicated vertical atmospheric structure that often exists at times of fog potential, a method for handling vertical stability variations may be required. This problem area, of course, is closely related to that of vertical diffusion from industrial stack plumes, and improved models would benefit air pollution areas as well as cooling tower concerns. Field data are needed to refine existing models.

Mechanical Draft Towers: It is generally accepted that mechanical draft towers present a much higher probability of ground level effects because of the height and nature of the discharge. Verified models for predicting the height of rise of mechanical draft plumes are needed. Effective plume rise from mechanical draft towers is a complicated problem because discharge takes place from towers consisting of lines of individual cells.

In turn, a number of towers may be grouped in various ways at large stations. Merging of individual plumes and the net plume rise effect from a complex of towers must be investigated. A satisfactory model should be capable of defining the relative effects of various tower numbers, spacings, and orientations. Aerodynamic effect of airflow over the towers themselves must probably be considered also in this problem.

Plume Model Applications and Extensions – Given satisfactory models for defining the behavior of plumes from a natural draft tower or a group of mechanical draft towers, there are several critical problems that should be attacked. These problems, in turn, should be the object of new model developments.

A first application, of fundamental importance for power plant design and siting, is delineation of the relative effects of wet and dry cooling towers, plants of various sizes, and various sizes and configurations of cooling towers. For example, it is not clear whether, for a given rate of energy dissipation, the plume from a dry tower in a humid climate is more likely to initiate clouds than the plume from a wet tower.

This model application and development must be carried out to determine whether new designs of towers can reduce atmospheric effects, and whether certain effects become critical for plants beyond a given size. Field data are needed before modeling is attempted.

The second related application is evaluation of weather influences on a larger scale than that of the plume. Here, the plume models can serve to define input or boundary conditions, and a different type of meteorological model will be required. Specific questions are:

When and in what conditions can a single cooling tower or a group of cooling towers initiate cloud formation over and downwind of the station?

To what extent will cooling tower effluents contribute to significant cloud development and ultimately to mesoscale precipitation?

On a climatological scale, how will cooling towers or groups of towers influence location and quantity of precipitation?

Is there a critical heat release rate for a given site below which no significant weather effects are produced and above which rainfall increases can be expected?

It seems quite obvious that there is a magnitude of local heat input to the atmosphere that must lead to significant changes in weather, particularly cloud formation and precipitation. The fundamental question, and perhaps the ultimate question to be answered by research into effects of thermal discharge, is the magnitude and space scales of heat input at which such significant effects will occur.

Model Development–Ponds and Reservoirs: The transfer of waste heat from cooling ponds, reservoirs, and spray ponds or canals requires mathematical modeling since there are almost no observational data to predict the possible problems. Also needed are better field data on the fluxes of latent, sensible, and radiant heat to and from cooling ponds and spray canals, and better computer simulations of these fluxes.

There is considerable disagreement in the literature on how to estimate these heat exchange rates; the published coefficients vary by a factor of at least 6 or 8. Better predictions of these fluxes will result in better designed cooling ponds (probably smaller) and less consumptive water losses. Since the transfer properties of the atmosphere are affected by the waste heat flux, the mathematical problem is nonlinear and is solved best by the use of numerical computer modeling.

As the waste heat diffuses through the atmosphere, large scale atmospheric phenomena are affected. For example, it is important to determine whether the waste heat from an intensive energy production center will influence synoptic conditions. The modeling techniques can help answer questions concerning the optimum size of power plants, the design of monitoring systems, and field research programs. For example, will a 10,000 megawatt plant on a 15,000 acre lake produce a significant weather modification in the mesoscale wind patterns and weather conditions?

Hydrologic

The objective of the physical and analytical studies of thermal discharges is to provide a predictive capability that will allow a priori estimates of the temperature and the spatial and temporal distribution of water discharged from power plants. These predictions must be sufficiently accurate to allow an estimate of the possible biological effects of the thermal discharge.

A number of mathematical models describing the dispersion of heated or buoyant discharges are available in the literature. Some of these models attempt to treat the entire discharge while others treat only the near field (jet momentum dominated) or the far field (drift flow). There have also been a number of physical scale modeling studies of either the near field or far field regimes.

Unfortunately, there has been practically no testing or validating of the models with actual field data. The greatest need in this area is the acquisition of adequate field data from operating power plants to allow the testing of both the mathematical and physical scale models.

In the momentum dominated near field region the major modeling difficulties are associated with the boundary conditions, such as the bottom effects, shoreline influences, ambient lake currents, and upwelling conditions, and their effects on the stratification of the heated water.

In the joining region, that which connects the near field and the far field regions, the water is usually stably stratified and diffuses primarily by horizontal spreading. None of the available models treat the often observed sharp interface on one side of the thermal plume that appears to be caused by gravity head spreading against the direction of the ambient lake current.

Dispersion in the far field region is the most difficult to model because of its dependence upon the ambient lake turbulence, wind induced surface currents, atmospheric heat transfer rates, and the transient nature of the discharge. To adequately predict the location of various areas of warm water the model must be a transient analysis. A great need exists for data on the ambient turbulence and eddy diffusivities in the water body. The greatest need, however, is a study of the biological significance of this far field region where temperatures are only 1° to 1½°C above ambient. If the region is not biologically significant, then the tremendous effort required to adequately model it will not be needed.

The art of physical scale modeling would be considerably enhanced if provisions were made for testing the modeling results with post-operational data from power plants. The acquisition of this type of data should be encouraged. Instrumentation systems, including simple instruments for measurement of ultra low velocity in the field, should be developed.

Cooling Ponds: The cooling pond is an alternative to once through cooling or cooling by use of a cooling tower. Originally, this concept was generally utilized to allow locating the generating station at or near the fuel source, but in recent years the legislative restrictions placed on once through cooling have further increased the importance of the cooling pond.

For definition purposes it is noted that the cooling pond differs from the lake in that it is a water body of comparatively limited acreage, relatively shallow depth, and usually man-made.

The technical literature reveals that some research has been conducted on natural and man-made lakes, but the practical application of this material indicates that additional research is required. The areas that require additional consideration are as follows:

(1) Improved techniques for designing and predicting the performance of cooling ponds, including techniques that consider or determine

   (a) The interaction of the near field and the far field effects.
   (b) The effects of wind speed on both the heat transfer mechanism and on the surface distribution of the heated water.
   (c) The effects of the distribution of heated water across the surface of the pond due to differences in water density resulting from temperature differences.
   (d) The significance of possible channeling of the heated water from the discharge source to the intake.

(e) The interaction of multiple separated heat sources on a common cooling pond

(f) The total effective surface area of the pond as a result of the consideration of (c) and (d) above. This item also refers to the determination of the effective cooling area resulting from adjacent upstream and downstream surface areas not in the primary flow pattern and considers inflow and outflow to the total pond as an additional variable. Additionally, the determination of the equivalent effective area of inlets, bays, etc., outside of the main flow channel should be considered.

(g) The influence of water depth on cooling pond performance and methods for maximizing benefits of thermal storage capacity in ponds.

(2) Improved analytical models which consider hourly weather, solar, and plant heat loads as well as other variables considered in (1), from which time history plots of all areas of the pond can be developed.

(3) Improved analytical techniques to determine the natural and forced evaporative water losses associated with ponds in conjunction with the development of (2).

(4) Methods to increase the efficiency of cooling ponds and to reduce the solar radiation by, for example, the introduction of reflective material on the pond surface.

(5) Complete and meaningful field data to verify the analytical models or to obtain empirical coefficients that can be applied to the analytical equations.

Spray Ponds/Canals: Spray canals and spray ponds of a large size are relatively new in their application to the dissipation of waste heat. Much has yet to be learned about their design and operation. The following areas need investigation.

Spray Pond/Canal Configuration – Various design configurations should be analyzed to minimize interference with surrounding structures and natural boundaries. Optimal thermal efficiency with respect to hydraulic, topographic and meteorologic conditions should be investigated. It may be beneficial to use long narrow canals/ponds laid out in a curvilinear pattern, to best accomplish this.

Spray Placement – Spray placement with respect to surrounding sprays is of primary importance, since degradation of efficiency of succeeding downwind nozzles must be considered in the design. Investigation of the degree of this degradation is necessary with the objective of optimizing nozzle density over the pond area.

Spray Design – Spray nozzle design parameters which maximize heat transfer per unit of cost and minimize drift problems by controlling drop size and area covered should be established.

Drift Control – The amount of drift and the distance it travels downwind should be investigated under various conditions of wind velocity, nozzle height, water pressure, and nozzle type. Various control methods for drift should be investigated, such as drift fences and drop size control devices.

Analytical Models – Both thermal and atmospheric models are necessary to establish boundaries on the effects of heat and moisture rejected by this method. Field correlation is necessary. The extent of downwind drift, dew, fog and icing needs to be established under various atmospheric conditions for the various regions of the country.

Technology and Engineering

In 1970, there were some 340,000 megawatts of installed electrical generating capacity in the contiguous United States. Of a total of about 3,400 facilities, about 2,400 are either hydroelectric or steam turbine/internal combustion units which provide relatively small quantities of overall power supply and either release no heat or release heat directly to the atmosphere.

The remaining 1,000 facilities are steam electric sources, requiring large inputs of thermal energy and releasing to the environment approximately two-thirds of their basic energy consumption. These generating systems require methods for carrying off excess quantities of thermal energy, normally accomplished by passing large quantities of water through the system and releasing this warmed water back to the environment, usually through streams and lakes.

There are many unanswered questions regarding the impact of waste heat on the environment, even at today's levels of water usage. Future electric utility installed capacities are predicted to reach 600,000 megawatts in 1980 and 5,200,000 megawatts in 2050. All of this capacity will require some method for removal of the excess waste heat, with the system so designed as to have minimal effects on the environment, in order to alter it as little as possible. This section of the book contains the conclusions of the group concerning the engineering and technological areas of the problem and is organized in three broad topical areas: improvements in existing technology; energy system analysis; and improved techniques for power generation.

Improvements in Existing Technology: Evaporative Cooling Towers – It is believed that the performance of cooling towers has not yet been fully exploited. Research is needed to reduce the plume or fog from mechanical and natural draft towers.

The presence of plume or fog is objectionable for aesthetic reasons, as well as for its production of high humidity, poor visibility, icing of surrounding property, etc. Suggestions for future research are as follows:

Investigation of electrostatic eliminators.

Investigation of after-heating and the effectiveness of the combined wet-dry tower.

Investigation of the effect of seeding.

Investigation of the variation in the shape, design, and baffling of the air intakes and outlets (for example, to develop a vertical plume by tangential air inlets).

Investigation of the effect of intermittent release of air discharge (puffing, torus-shaped cloud discharge, etc.).

Investigation of the effect of stainless steel wire mesh, plastic particulate parts, etc., for control of drift.

Research is needed to develop a reliable method for measuring drift from a mechanical or natural draft cooling tower. It is reported that a laser beam scattering method is being used, but information is incomplete on its applicability and accuracy.

Research is needed to reduce drift, particularly from wet mechanical draft cooling towers, but also from natural draft towers. Present guarantees for cooling towers call for less than 0.2% of the tower circulating rate, but this is sufficient to cause appreciable deposits on surrounding objects. Baffles in which the discharge airflow is changed in direction are now used, but other means such as stainless steel mesh, different baffle design, plastic particulate shapes, etc. should be investigated.

Dry Cooling Towers – For some time dry cooling has had an application in the petroleum and petrochemical industry to disperse waste process heat where the temperature reaches 200° to 400°F. It is only recently that there has been application in the power industry where the temperature reaches 80° to 100°F. The power industry has some major obstacles to overcome before dry cooling systems come into widespread use.

An obstacle will be the cost of large steam turbines which have yet to be designed for the high back pressure. Dry cooling tower structures are much more expensive than other methods of cooling, at least when the final temperature must be as low as possible.

Research to reduce this economic disadvantage is necessary. Deserving of investigation are areas such as improved fin design and materials, roughness treatment, transition and turbulence control, use of swirl flow, liquid additives, suspensions, vibration, electrostatic fields, fluid injection, and suction/blowing boundary layer controls.

Although the indirect system using an intermediate heat transfer fluid seems best suited for large plants, further attention should be given to the system which condenses the turbine exhaust by direct cooling with air. This involves some rather difficult fluid handling problems due to the low pressure and high specific volume of the turbine exhaust.

Not to be overlooked is the significance of establishing large heat islands and their effect on the environment. Finally, consideration of the effect of dry cooling on the total available energy picture must be ascertained. In most United States climates, dry cooling results in poorer efficiencies and therefore greater fuel usages.

Combined Cooling Methods – The needed techniques to determine the individual performance and physical effects on the environment of various cooling methods have been discussed. There is also a need to investigate the possible advantages of combining cooling methods, and, if there indeed are advantages, to develop methods to determine optimum combinations.

It must be recognized that regional considerations would be an important variable in these investigations. A typical combination that might be considered is the use of spray channels in combination with a cooling pond. The spray channel would most likely be operated only when peak generating loads and waste heat output are the largest. Continued investigation is also suggested in the following areas:

Combined wet and dry towers
Dry towers with spray cooling of the inlet air
Fan assisted dry and wet natural draft towers
Dry tower and a cooling pond

Other Waste Heat Transfer Methods – Possible alternatives to lakes, ponds, towers, etc. which have not yet been invented would be welcome innovations. These would have to be analyzed from both an engineering and economic basis. However, conceptual studies including order of magnitude considerations might be useful.

Evaluation and development of once through cooling systems should be continued in order to seek out the best possible solutions to the problems and questions of environmental waste heat. New methods for improving the performance of once through cooling systems, such as new techniques of dispersal or tailoring of plumes and jets, should be pursued.

Underground water sources and sinks could provide a waste heat transfer system. This must be considered from both a water source point of view and a point of view toward ground water stability. Technically, there is a severe heat transfer problem regarding thermal ground loading. Unfortunately, due to the large quantities of water involved, this aspect has a very low priority.

Beneficial Uses of Waste Heat – Heated fluids discarded at 80° to 100°F in summer and 45° to 55°F in winter do not have many uses, and are expensive to transport any distance to a point of use. Power stations can reject heat at higher levels (300° to 380°F) for useful purposes, as for example, urban heating and cooling, but the electrical generation of the power station, for which purpose the plant was constructed, is significantly reduced. Two definite advantages in doing this are in cost sharing and in overall conservation of natural resources.

More efficient sewage treatment may be attained by applying waste heat in this process, since an 18°F rise in temperature approximately doubles the biochemical reaction rate. Final treated sewage effluent, which has been treated for phosphate removal, may also be employed satisfactorily for cooling condenser water in a cooling tower, as has been done at Amarillo, Texas, over the past 10 years.

A better quality effluent in respect to organic decomposition and increased oxygen content should be discharged to the stream as a result of aeration provided in the cooling tower.

Investigations are needed on other beneficial uses, such as using warm water in the process industries, in greenhouses, in aquaculture to increase yield, in irrigation of field crops, and in a variety of heating applications (soils, highways, etc.). Although the economic problem associated with transport and use of large volumes of warm water are formidable, continued attention to such concepts is warranted in order to fully utilize every bit of our energy.

Energy Systems Analysis: The increasing emphasis on environmental impact of power plant waste heat discharges requires the establishment of planning and design guidelines with analytical techniques for evaluation and decision making. This must go beyond optimization of individual components and determination of effects.

Evaluating the consequence of particular directives, alternatives, and the procedure of optimization are embodied within systems analysis and operations research. It is felt that much can be done in energy systems approach along this line. The approach should include technical, economic, ecological and sociological factors as they apply to system identification, modeling, optimization and evaluation.

Improved Techniques for Power Generation: Modern fossil-fuel steam electric generating plants operate at approximately 32 to 38% efficiency in converting fuel to electrical energy, and most nuclear-fuel plants operate at about 32%.

Thus, about two-thirds, of the total energy expended is not converted to electrical energy and must be dissipated to the environment. This section indicates areas of research and development needed to increase the efficiencies of power production in order to reduce any further degradation.

Improved Efficiencies of Existing Systems — Improvement in the efficiencies of fossil-fuel generating systems over the present percentages would reduce consumption of dwindling natural resources, and decrease the amounts of waste thermal energy. Increases in the efficiencies of nuclear-fuel steam supply systems from the present 32% to about 40% through the use of gas-cooled or breeder reactors would also reduce the amounts of waste heat to the environment.

Development of the Breeder Reactor Systems – Breeder reactor systems, which essentially produce as much fission fuel (plutonium-239 or uranium-233) as is consumed (uranium-235), will provide a vitally needed stopgap measure to conserve natural resources of nuclear fuels, while other power sources are being developed. In addition to the conservation of fuel, these reactors may operate at considerably higher efficiencies than are typical today, thereby releasing less waste heat to the environment.

Development of Open Cycle–Combustion Magnetohydrodynamic Generator Systems – In the magnetohydrodynamic (MHD) concept, the energy contained in a very hot stream of electrically conducting gas is converted directly into electrical energy. A small amount of seed material, such as cesium or potassium carbonate, must be injected into the combustion chamber to make the gas stream a good conductor.

The charged particles of the heated gases pass through a magnetic field, generating an electrical current. Efficiencies in this system may increase significantly, substantially lowering the amount of waste heat which must be discharged. In addition, topping and/or bottoming cycles can be considered to obtain greater efficiencies.

Much research and development are needed to make this a practical system, and it is important that this be done. In addition to utilizing the higher temperatures for improved thermal efficiency, MHD may also allow a more favorable fuel choice. The open cycle system may be able to use fuels now unsuitable for air pollution reasons. Environmental effects should be investigated simultaneously.

Development of Closed Cycle–Nuclear Magnetohydrodynamic Generator Systems – The increasing use of nuclear energy suggests the significance of a gas-cooled reactor with an MHD device in a closed loop. However, the lower temperatures of the current nuclear devices suggest that a nonequilibrium operation is essential. Here, internal electrical currents elevate electron temperature of the seed material ionization.

The nonequilibrium is maintained through ineffective collisions of particles. Both MHD systems are subject to important problems such as end and surface effects, instability, and materials degradation, which all require continuing research.

Development of Fusion Reactor Systems – Fusion, whereby the nuclei of light elements fuse to form a heavier element and also release energy, could ultimately solve the problem of natural resources and scarcity of fuels. Efficiencies are potentially the greatest of any thermal system because of the high temperatures involved. The most significant problem at present is fusion plasma containment.

Development of Solar Energy Utilization – The impingement of solar energy on the surface of the earth is a massive energy source yet untapped. Development of economical methods to harness this energy will provide a complete combustion-free system which will not add any extra thermal energy to the environment, but will merely delay its reradiation while also directing it into productive usages.

Development should be directed toward small-scale use on a household level, and toward central power station use as a supplement to major power sources.

Development of Methods for Flattening Peaks of Energy Usage – Development of items such as rechargeable fuel cells in the house which could be recharged at off peak hours (overnight) would allow for a slower growth rate of power sources and would provide more efficient utilization of available facilities.

Methods for reducing consumption of electrical power during prime times may be developed by such means as premium billing rates, storage facilities for off peak energy, storage of cold refrigerant, etc.

Development of Improved Transmission and Dispatching Techniques – Development of improved methods for electrical power transmission, particularly over long distances, and improved reliability of transmission facilities, will reduce the attendant heat load rejected to the environment.

Further, the initiation of a national integrated network of power dispatching will further reduce the number of generating facilities required, and will allow generating facilities to be operated at a higher fraction of their capacity.

ENERGY IN THE CITY ENVIRONMENT 1973

Edited by Robert N. Rickles

This book is based upon material supplied by Robert Rickles, the Institute for Public Transportation, and the New York Board of Trade.

The important **charts** in the **appendix** present a list of energy conservation measures for the short-term (1972-1975), mid-term (1976-1980) and the long-term (beyond 1980) for the Transportation, Residential/Commercial, Industrial and Electric Utility sectors. The charts also indicate estimated maximum attainable energy savings, possible means for implementing each conservation measure, and pros and cons.

ISBN 0-8155-5019-7 **173 pages**

ENERGY FROM SOLID WASTE 1974

by Frederick R. Jackson

Pollution Technology Review No. 8

Energy Technology Review No. 1

The solid waste disposal problem is reaching alarming proportions everywhere. The United States alone produces close to 300 million tons of solid waste per year, which is equivalent to about one ton per person.

At the present time the prevalent methods of disposal are dumping and sanitary landfill. Municipal incineration disposes of a small portion only, with attendant high capital and operating costs.

Many methods have been proposed for coping with the problem, such as source separation, source reduction, or material recovery. However, with the energy crisis descending upon us, producing energy from waste is becoming more and more attractive. An estimate currently making the rounds in financial circles is that when the price of crude oil reaches $7.00 a barrel, alternate sources of energy become practicable.

This book is based primarily upon information from studies conducted under the auspices of the EPA. Its foremost topic is burning of solid wastes to create steam directly. The air pollution problem created by burning can be solved quite easily with known technology. Solid wastes are low in sulfur, consequently there is no SO_2 removal problem.

Another technique that may assume more importance in the future is the controlled pyrolysis of wastes, yielding so-called pyrolysis gas or oil. Chapter seven is devoted to this. The final chapter discusses European practice, which is historically far more extensive than that of the U.S. A condensed table of contents follows here:

ISBN 0-8155-0528-0 **163 pages**

WASTEWATER CLEANUP EQUIPMENT 1973

Second Edition

Water pollution is becoming more of a problem with every passing year. Plants engaged in all types of manufacture are being more and more carefully watched by federal, state, and municipal governments to prevent them from pouring their untreated effluents into the nation's waterways, as they used to do. The sewage treatment plants of many municipalities are becoming too small for the burgeoning population, and many communities once served by individual septic tanks are having to build sewers and treatment plants.

Water pollution will be solved primarily by application of techniques, processes, and devices already known or in existence today, supplemented by modifications of these known methods based on advanced technology. This book gives you basic technical information and specifications pertaining to commercial equipment currently available from equipment manufacturers. Altogether the products of 94 companies are represented.

This second edition of "Wastewater Cleanup Equipment" supplies technical data, diagrams, pictures, specifications and other information on commercial equipment useful in water pollution control and sewage treatment. The data appearing in this book were selected by the publisher from each manufacturer's literature at no cost to, nor influence from, the manufacturers of the equipment.

It is expected that vast sums will be spent in the United States during the remaining portion of this decade for control and abatement of water pollution. Much of the expenditure will be for the type of equipment described in this book.

Today's environmental control is taken to mean a specialized technology employing specialized equipment designed to process the discarded and excreted wastes of human metabolism and human activity of any sort.

Next to air, water is the most abundant and utilized commodity necessary for the maintenance of human life. The average consumption of water per person in residential communities in the United States is between 40 and 100 gallons in one day. In highly industrialized communities the average consumption pro head can be as high as 250 gallons per day.

The reuse of wastewater after cleanup is not only becoming a cogent necessity, but it is also becoming more attractive economically. The degree of purity required for industrial water use is in many cases greater or vastly different from that acceptable for potable water.

Special equipment for cleanup of wastewater is therefore an absolute must, and this book is offered with the intention of providing real help in the selection of the proper equipment.

The descriptions and illustrations given by the original equipment manufacturer include one or more of the following:

1. **Diagrams of commercial equipment with descriptions of components.**
2. **A technical description of the apparatus and the processes involved in its use.**
3. **Specifications of the apparatus, including dimensions, capacities, etc.**
4. **Examples of practical applications.**
5. **Graphs relating to the various parameters involved.**

Arrangement is alphabetically by manufacturer. A detailed subject index by type of equipment is included, as well as a company name cross reference index.

ISBN 0-8155-0487-X **372 pages**

ENVIRONMENTAL LITERATURE 1973

A Bibliography

by Gary F. and Judith C. Bennett

This book deals with all aspects of pollution and pollution control, it being a bibliography of books, pamphlets, journals (but not of journal articles), and of some audiovisual aids in a field that is now euphemistically named "environmental."

This pollution bibliography is actually the third edition of this work and is about twice the length of the preceding edition, published in June 1969 by the University of Toledo, Ohio under the title, "Bibliography of Books on the Environment: Air, Water and Solid Wastes."

Since that time there has been a substantial increase in the number of primary sources and publications, i.e. journals, books, technical reports, symposia, academic literature and government documents, both domestic and foreign, that contain information about the environment and related subjects.

As in all fields pertaining to the Life Sciences the volume of literature with respect to the number of sources and diversity of contents has led to selective dissemination of information (SDI) on most aspects of environmental information.

This in turn has led to the generation of a multitude of secondary publications (abstracts, excerpts, summaries, card services, "trailing microfiche" publications and many more). The secondary publication is the SDI link for a user's specific information need as he confronts thousands of primary information titles and choices.

Because they fill a practical need, and afford ease of use and reference, such secondary publications have been listed meticulously in this edition.

Title listings of primary publications are complete and are never abbreviated, thus giving a clear indication of what information will be found in the original publication cited. Users of this book have a right to know which references deserve requests for photocopies or inter-library loan. Emphasis is on American publications with a good selection of foreign titles.

Another special merit of this volume is the inclusion of abstracts of meetings and symposium proceedings—all titles which are hard to come by and usually present an arduous and time-consuming task for the librarian. Meeting papers are often changed or augmented considerably, while getting them ready for the printer. Persistent efforts by the compilers in order to verify such information have contributed in a large measure to insure this book's high accuracy and utility.

The literature is divided into categories. It is realized that this division is often arbitrary, for publications that comprise more than one topic are placed into what constitutes the most suitable category in the authors' estimation. The senior author is well qualified to make such selections, he being a Professor of Biochemical Engineering at the University of Toledo, and Chairman of A.I.Ch.E.'s Environmental Division.

To give the prospective user an approximation of the wealth of information available in this book, the table of contents is reproduced here:

Table of Contents

ISBN 0-8155-0509-4 **134 pages**

POLLUTION DETECTION AND MONITORING HANDBOOK 1974

by Marshall Sittig

Environmental Technology Handbook No. 1

This handbook contains methods for the detection and monitoring of pollutants in industrial effluents, notably air and water.

Such methods are prerequisites for any kind of management of environmental quality. Of equal importance are the data handling systems that must be used to collect and interpret the measurements, regardless of whether manual or automated identifying and monitoring techniques are chosen.

Generally the most important single factor in determining whether manual or automated methods of analysis and level monitoring should be used, is the required frequency interval. In the manual approach costs vary almost directly with the measurement frequency. At monitoring intervals oftener than once a day, or when a continuous watch must be maintained, installation of automatic sensing is more economical, provided reliable automatic sensors are available.

For meaningful control and prevention of pollution the accepted standards and parameters for air and water quality must be known, whether they have been legislated or not. Acceptable levels for each type of impurity from methylmercury to dissolved chlorides or carbon particles must be known by the industrialist or public health official, and this book, with its 1633 references makes a massive attempt to communicate such levels.

As a guideline for the industrial user, data on the toxicity of the various pollutants are included. Such data on the deleterious effects were taken from various government publications. This book is not a treatise of pharmacology or industrial toxicology, but a guide to the accurate evaluation of pollutant levels in a given effluent or ambient substrate.

Whenever applicable, the name of the pollutant is followed by chemical and other information. After this come directions for sampling in air or water or both, as the case may be. Thereafter are given qualitative and quantitative analytical methods suitable for identification and measurement of the degree or level of pollution. This comprehensive information is followed by measurement techniques suitable for repeated or continuous monitoring. Preference is given to methods giving reproducible results in environmental quality control, as recommended by the EPA and other U.S. government agencies. This is followed in each case by many up-to-date references to government publications, patents, and journal articles.

Arrangement is encyclopedic. The book is thus a worthy companion to the author's POLLUTANT REMOVAL HANDBOOK available from the same publisher.

Pollution control is a serious business, and in the design of a monitoring program the specific objectives to be served must be clearly in mind from the very beginning. Otherwise the effort most probably will not result in an efficient and meaningful program. The length of the initial survey, selection of monitoring points, parameter coverage, and sampling frequency all depend on the specific objectives and the timeframe in which the objectives must be achieved. These variables also have a very significant impact on monitoring costs.

Monitoring for the purpose of long-term trend identification, evaluation of standards compliance, and total management of environmental quality must, of course, be a continuing program without end.

It is hoped sincerely that this book, together with its companion volume POLLUTANT REMOVAL HANDBOOK, will do much to establish and sustain such antipollution programs.

A partial and condensed table of contents follows here in which subheadings are indicated below for only the first few entries. The book contains a total of 88 subject entries arranged in an alphabetical and encyclopedic fashion. The subject name refers to the polluting substance, and the text underneath each entry tells how to measure and monitor pollution by said substance:

POLLUTION INSTRUMENTATION
Size of the Market
References

SAMPLING
Air and Gas Sampling
Water Sampling
References

SUBSTANCES AND MATERIALS

ACIDS AND ALKALIS

ALDEHYDES

ALUMINUM

AMMONIA

ANTIMONY

ARSENIC

ASBESTOS

AUTOMOTIVE EXHAUST

BACTERIA

BARIUM

BERYLLIUM

BISMUTH

BORON

BROMIDES

CADMIUM

CALCIUM

CARBOHYDRATES

CARBON

CARBON DIOXIDE

CARBON MONOXIDE

CHLORIDES

CHLORINATED HYDROCARBONS

CHLORINE

CHROMIUM

COBALT

COPPER

CYANIDES

ETHYLENE

FLUORIDES

HALOGENS

HALOGEN COMPOUNDS

HYDROCARBONS

HYDROGEN CHLORIDE

HYDROGEN SULFIDE

IODIDES

IRON

LEAD

LEAD ALKYLS

MAGNESIUM

MANGANESE

MERCAPTANS

MERCURY

MOLYBDENUM

NICKEL

NICKEL CARBONYL

NITRATES

NITRIC ACID

NITRITES

NITROGEN COMPOUNDS

NITROGEN OXIDES

ODOROUS COMPOUNDS
Measurement in Air
Sampling Methods
Qualitative Analyses
Quantitative Methods (Organoleptic)
Quantitative Methods (Instrumental)
Measurement in Water
References

OIL AND GREASE

ORGANICS

OXYGEN DEMAND

OZONE AND OXIDANTS

PARTICULATES

PEROXYACETYL NITRATE (PAN)

PESTICIDES

PHENOLS

PHOSGENE

PHOSPHATES

PHOSPHITES

PHOSPHORIC ACID

PHOSPHORUS

PHOSPHORUS COMPOUNDS (ORG.)

POLLEN

POLYNUCLEAR AROMATICS

POTASSIUM

RADIOACTIVE MATERIALS

SEDIMENTS

SELENIUM

SILVER

SODIUM

STRONTIUM

SULFATES

SULFIDES

SULFITES

SULFUR COMPOUNDS (ORG.)

SULFUR DIOXIDE

SULFUR TRIOXIDE

SULFURIC ACID

SURFACTANTS

TIN

TITANIUM

URANIUM

VANADIUM

ZINC

ZIRCONIUM

FUTURE TRENDS

This book should prove to be very useful as an overall reference volume. The organization is clear, and the text is especially well arranged. The tables and figures are extensive and will be of value to students, engineers, and pollution control authorities.

ISBN 0-8155-0529-9

401 pages

POLLUTION CONTROL IN THE ORGANIC CHEMICAL INDUSTRY 1974

by Marshall Sittig

Pollution Technology Review No. 9

Wastes from plants manufacturing identical compounds may be quite dissimilar, because of the difference in processes and raw materials. Since the organic chemical industry is now largely petrochemical, preference has been given to waste treatment from such operations.

Detailed treatability evaluation of each waste stream has become a prerequisite for any profitable product.

Physical methods for waste and by-product removal include gravity separation, air flotation, filtration, evaporation and adsorption techniques.

Chemical methods include precipitation, polyelectrolyte treatment, oxidation and other chemical conditioning such as neutralization.

Biological treatment methods comprise activated sludge and its modifications, trickling filters, aerated lagoons, and waste stabilization ponds.

This book intends to assist the chemical engineer in treating chemical waste products and effluents in conjunction with prudent raw material choices and thus bring about process economics. It presents condensed vital data from government and other sources of information that are scattered and difficult to pull together.

A partial and condensed table of contents follows here.

Dinitrotoluene from Nitrotoluene

Disulfoton from Ethanol

Dyes and Pigments from Aromatics

Epichlorohydrin from Allyl Chloride

Ethylbenzene from Benzene

Ethyl Chloride by
Hydrochlorination of Ethylene
Chlorination of Ethane
Hydrochlorination of Ethanol

Ethylene or Propylene from Paraffins

Ethylene Dichloride from Ethylene

Ethylene Glycol from Ethylene Oxide

Ethylene Oxide from Ethylene

Formaldehyde from Methanol

Long Chain Alcohols from
Ethylene Oligomers

Malathion® from Methanol

Methanol from Natural Gas

Methylamines from Methanol + Ammonia

Methyl Bromide from Methanol

Methyl Methacrylate from
Acetone Cyanohydrin

Nitrobenzene from Benzene

Nitrochlorobenzene from Chlorobenzene

Nitroparaffins from Paraffins

Oxo Products from Olefins

Parathion from p-Nitrophenol

Perchloroethylene from Propane

Phenol from Chlorobenzene

Phenol from Cumene

Phorate from Ethanol

Phosgene from CO + Cl_2

Phthalic Anhydride from
Naphthalene or Xylene

Propylene Glycol from Propylene Oxide

Propylene Oxide from Propylene

Styrene from Ethylbenzene

SNG from Crude Oil

2,4,5-T from Trichlorophenol

Terephthalic Acid from Xylene

Tetraethyl Lead from Ethyl Chloride

Trichloroethane from Vinyl Chloride

Trichloroethylene by
Chlorination, then by
Dehydrochlorination of Acetylene
or by Oxyhydrochlorination
of Dichloroethane

Trifluralin from Perfluoromethylchloro-benzene

Vinyl Acetate from
Ethylene and Acetic Acid

Vinyl Chloride from Acetylene

Vinyl Chloride from Ethylene Dichloride

AIR POLLUTION CONTROL
Halogen Acids
Halogenated Hydrocarbons
Halogens
Hydrocarbons
Particulates
Sulfur Compounds
Coking of Coal

WATER POLLUTION CONTROL
Primary
Oil Separation
Equalization
Neutralization
Sedimentation
Flotation
Flocculation
Nutrient Addition
Secondary
Activated Sludge
Extended Aeration
Trickling Filters
Aerated Lagoons
Waste Stabilization Ponds
Chemical Oxidation
Nitrification—Denitrification

Tertiary
Chemical Precipitation
Gas Stripping
Microstraining
Carbon Adsorption
Electrodialysis
Ion Exchange
Evaporation
Reverse Osmosis
Chlorination
Rapid Sand Filtration
Sludge Handling
Aerobic Digestion
Anaerobic Digestion
Wet Oxidation
Thickening
Lagooning
Sand Drying Beds
Vacuum Filtration
Filtration
Land Disposal
Incineration
Sea Disposal
Ultimate Disposal
Thermal Oxidation
Deep Well Disposal

OVERALL WATER POLLUTION CONTROL MODELS
BPCTCA: Best Practicable Control Technology Currently Available
BATEA: Best Available Technology Economically Available
BADCT: Best Available Demonstrated Control Technology

THE ECONOMICS OF POLLUTION CONTROL

FUTURE TRENDS

ISBN 0-8155-0536-1

305 pages

SANITARY LANDFILL TECHNOLOGY 1974

by Samuel Weiss

Pollution Technology Review No. 10

This Pollution Technology Control Review surveys all the latest technical information on this subject. It is based primarily on studies conducted by industrial or engineering firms, under the auspices of the Environmental Protection Agency.

Sanitary landfill practice is an engineering method of disposing of solid wastes on land, whether they are of municipal or industrial origin. There is no resemblance to the old-fashioned garbage and rubbish dump. There are no fires, no obnoxious fumes or smoke, no flies and no rodents or other scavengers.

As stated in the introduction to this book, a sanitary landfill is not only an acceptable and economic method of solid waste disposal, it provides also an excellent way to improve the commercial value of otherwise unsuitable or marginal land areas within a few years.

A partial and condensed table of contents follows here.

ISBN 0-8155-0542-6 **300 pages**

LARGE SCALE COMPOSTING 1974

by M. J. Satriana

Pollution Technology Review No. 12

Composting is one of the oldest solid waste disposal methods known to man. This microbiological treatment process has the capability of converting municipal refuse and other organic waste solids into a product which has a lower bulk than the original waste, is stable, and can be recycled into the ecological sphere.

Aerobic, thermophilic composting is generally believed to provide the most rapid and complete decomposition of oxidizable or putrescible organic matter. It is especially applicable to decomposing material consisting primarily of discrete organic solids with open pore spaces.

Rough-quality compost, cheaply produced, has real potential for the reclamation of poor and spoiled lands (as from strip mining) providing it can be transported there cheaply (e.g. by barges).

This book, based primarily upon reports produced under the auspices of the Environmental Protection Agency, presents an overview existing large-scale composting methods, both here and abroad. It compares costs and partial recovery of these costs, but retains a significant emphasis on the workable disposal of urban waste versus other and usually more expensive methods of getting rid of solid organic refuse. Nine chapters. A partial and condensed table of contents follows here.

ISBN 0-8155-0549-3 **269 pages**